P9-DFC-018

Application of Green's Functions in Science and Engineering

PRENTICE-HALL, INC. ENGLEWOOD CLIFFS, NEW JERSEY

Application of Green's Functions in Science and Engineering

Michael D. Greenberg
Department of Mechanical and Aerospace Engineering
University of Delaware

PRENTICE-HALL INTERNATIONAL, INC., *London*
PRENTICE-HALL OF AUSTRALIA, PTY. LTD., *Sydney*
PRENTICE-HALL OF CANADA, LTD., *Toronto*
PRENTICE-HALL OF INDIA PRIVATE LIMITED, *New Delhi*
PRENTICE-HALL OF JAPAN, INC., *Tokyo*

Current printing (last digit):
10 9 8 7 6 5 4 3 2 1

13-038836-X

Library of Congress Catalog Card No. 75-153445
Printed in the United States of America

To My Parents,
Abraham and Celia

Preface

It is often felt that the Green's function method is too sophisticated to be widely available to engineers and scientists. I have tried to show, here, that this reputation is unjustified.

Our discussion of the method is based largely upon the consideration of a wide variety of applications, some posed in a physical setting, and others in purely mathematical terms. The approach is somewhat formal, with little consideration of such questions as existence or uniqueness, for example.

Most examples are followed by a series of *Comments* which expand upon various points in the solution, or relate to the mathematical and physical interpretation of the results. They are by no means afterthoughts, and constitute an important part of the text. In fact, a good deal of information is also contained in the *Exercises*. We strongly recommend at least a quick reading of the *Exercises*, even if the reader does not intend to work them.

We have done our best to keep the material as self-contained as possible, so that it will be accessible to a wide range of readers. We believe that it should be suitable either for advanced undergraduates or graduate students in the physical sciences or engineering, with only a modest background required in ordinary and partial differential equations. Beyond this, we assume the reader is acquainted with Green's theorem, the divergence theorem, and Laplace and Fourier transforms—but not necessarily with Laplace inversion. Although complex variable methods are used briefly, those few pages can be omitted with no loss of continuity.

Aside from the development of the Green's function method, we have included a good deal of basic information, of a more general nature, on boundary value problems, generalized functions, eigenfunction expansions, partial differential equations, and acoustics.

Finally, I would like to thank Professor Jerry L. Kazdan (University of Pennsylvania) for a wide variety of helpful comments on both the original and final drafts, mostly on the mathematics, but also on the style of presentation; to Professors A. Brooks Harris (University of Pennsylvania) and Morton M. Denn (University of Delaware) for detailed criticism of the original draft; to Professors H. D. Block, Y. H. Pao, and Bruno A. Boley (Cornell University) for various suggestions along the way; to Mr. James M. Mann for catching several errors in the manuscript; and, finally, to Mrs. Bertha R. Hollopeter, Mrs. Dorothy E. Carroll, and my wife Mimi for bearing up under the tedious typing.

MICHAEL D. GREENBERG

Newark, Delaware

Contents

PART I
Application to Ordinary Differential Equations

PART II
Application to Partial Differential Equations

Application of Green's Functions in Science and Engineering

I. Application to Ordinary Differential Equations

The first three sections of PART I introduce basic concepts such as linearity, superposition, and adjoints, with somewhat more attention devoted to the delta function and generalized functions in Section 3. The development of the Green's function method is contained entirely in Section 4, and is followed by a brief discussion of eigenfunction methods in Section 5.

1. INTRODUCTION

In PART I we will be concerned with the solution of the ordinary differential equation

$$\mathbf{L}u = \phi \tag{1.1}$$

over an interval $a \leq x \leq b$, subject to certain boundary conditions, where $\mathbf{L}$ is an nth order linear ordinary differential operator. By **linear**, we mean $\mathbf{L}$ is such that

$$\mathbf{L}(\alpha v + \beta w) = \alpha \mathbf{L}v + \beta \mathbf{L}w \tag{1.2}$$

for arbitrary constants α, β and any functions v, w which are at least sufficiently differentiable for $\mathbf{L}v$ and $\mathbf{L}w$ to exist.[1] We state, without proof, that $\mathbf{L}$ must therefore be of the form

$$\mathbf{L} = a_0(x)\frac{d^n}{dx^n} + a_1(x)\frac{d^{n-1}}{dx^{n-1}} + \cdots + a_n(x) \tag{1.3}$$

It is easily verified that this does in fact satisfy the linearity requirement (1.2).

By contrast, let us consider a differential operator which is *not* of the form (1.3); for example,

$$\mathbf{N}u = u'' - (u')^2 + 5xu \tag{1.4}$$

where we denote $d(\quad)/dx = (\quad)'$ for brevity. Applying $\mathbf{N}$ to an arbitrary linear combination $\alpha v + \beta w$ we have

$$\begin{aligned}\mathbf{N}(\alpha v + \beta w) &= (\alpha v'' + \beta w'') - (\alpha v' + \beta w')^2 + 5x(\alpha v + \beta w)\\ &= \alpha[v'' - (v')^2 + 5xv] + \beta[w'' - (w')^2 + 5xw]\\ &\quad - (\alpha v' + \beta w')^2 + \alpha(v')^2 + \beta(w')^2\\ &= \alpha \mathbf{N}v + \beta \mathbf{N}w + [\alpha(v')^2 + \beta(w')^2 - (\alpha v' + \beta w')^2]\end{aligned} \tag{1.5}$$

Clearly, the quantity inside the last square brackets is not identically zero for all allowable α's, β's, v's and w's, so that $\mathbf{N}$ does not satisfy the fundamental requirement (1.2), and is therefore *nonlinear*. This, of course, is no great surprise since $\mathbf{N}$ is clearly not of the form (1.3).

Now let us consider the *boundary conditions* associated with (1.1). Since $\mathbf{L}$ is of nth order, there will be n such conditions of the general form[2]

[1] We note that *all quantities will be understood to be real-valued*. Exceptions to this, as occur in the use of integral transforms and conformal mapping, will be apparent.

[2] Occasionally one must deal with boundary conditions of other types. For example, one such boundary condition might consist of the requirement that u be *finite* at one of the endpoints.

$$\mathbf{B}_j(u) = c_j; \qquad j = 1, 2, \cdots, n \tag{1.6}$$

where the c_j's are given constants, and the $\mathbf{B}_j$'s are prescribed functionals[3] of the unknown u. More specifically, *we will limit our attention throughout to* $\mathbf{B}_j$*'s which are linear combinations of* u *and its derivatives, through order* $n - 1$, *at the two endpoints* a, b. For $n = 2$, for example, we have

$$\begin{aligned} \mathbf{B}_1(u) &= a_{10}u(\mathrm{a}) + a_{11}u'(\mathrm{a}) + b_{10}u(\mathrm{b}) + b_{11}u'(\mathrm{b}) = c_1 \\ \mathbf{B}_2(u) &= a_{20}u(\mathrm{a}) + a_{21}u'(\mathrm{a}) + b_{20}u(\mathrm{b}) + b_{21}u'(\mathrm{b}) = c_2 \end{aligned} \tag{1.7}$$

We say that they are *linear* functionals, because

$$\begin{aligned} \mathbf{B}_j(\alpha v + \beta w) &= a_{j0}[\alpha v(\mathrm{a}) + \beta w(\mathrm{a})] + a_{j1}[\alpha v'(\mathrm{a}) + \beta w'(\mathrm{a})] \\ &\quad + b_{j0}[\alpha v(\mathrm{b}) + \beta w(\mathrm{b})] + b_{j1}[\alpha v'(\mathrm{b}) + \beta w'(\mathrm{b})] \\ &= \alpha[a_{j0}v(\mathrm{a}) + a_{j1}v'(\mathrm{a}) + b_{j0}v(\mathrm{b}) + b_{j1}v'(\mathrm{b})] \\ &\quad + \beta[a_{j0}w(\mathrm{a}) + a_{j1}w'(\mathrm{a}) + b_{j0}w(\mathrm{b}) + b_{j1}w'(\mathrm{b})] \\ &= \alpha\mathbf{B}_j(v) + \beta\mathbf{B}_j(w) \end{aligned} \tag{1.8}$$

for arbitrary constants α, β and functions v, w.

Now, thus far we have discussed both the differential operator $\mathbf{L}$, and the boundary conditions $\mathbf{B}_j$ which determine the domain of $\mathbf{L}$ and hence complete the specification of the operator. (Note that we distinguish between the term **differential operator,** which refers to $\mathbf{L}$ alone, and the term **operator,** $\mathscr{L}$ say,

[3] To clarify what is meant by the term **functional,** let us contrast it with the classical notion of a **function,** say $f(x)$. We say that to each x in the *domain* of f (i. e., the set of all x's for which f is defined; e. g., an interval a, b of an x axis) the function f assigns a numerical value $f(x)$ or, equivalently, a point on an $f(x)$ axis. The totality of $f(x)$ points is called the *range* of f; see Fig. 1. 1. Whereas the symbols f and $f(x)$ are generally used interchange-

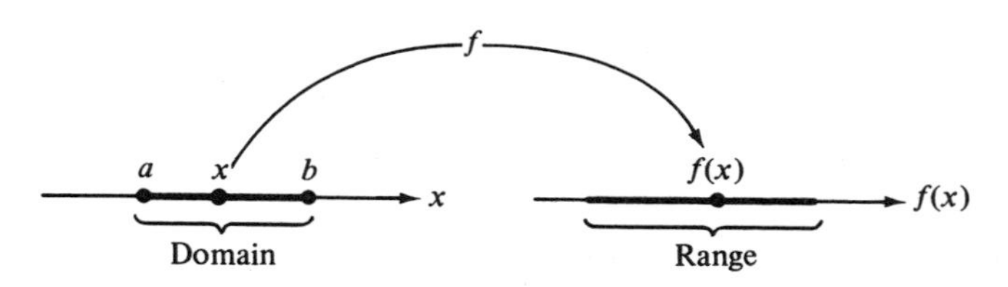

FIGURE 1.1 Domain and range of a function f.

ably, with no resulting confusion, it is worth mentioning that there is a difference, and occasions do arise where it is important to respect this difference in the interest of clarity. Specifically, $f(x)$ actually denotes the range point associated with the domain point x, whereas the single letter f denotes the entire mapping, as can be displayed graphically if, following Descartes, we arrange the x and $f(x)$ axes at right angles (hence the term which refers to **L** *plus* the $\mathbf{B}_j$ boundary conditions.[4]) We have been careful to require that both **L** and the $\mathbf{B}_j$'s be linear. This is of crucial importance here, since it implies the validity of **superposition,** which will be basic to the Green's function method.

To illustrate the idea of superposition, consider, for example, the second order system

$$xu'' + u = \phi; \qquad \begin{aligned} u(0) &= 3u(1) \\ u'(0) &= 0 \end{aligned} \tag{1.9}$$

over the interval $0 \leq x \leq 1$. First, let us rewrite this in the form of (1.1), (1.7):

$$\mathbf{L}u = xu'' + u = \phi \tag{1.10a}$$

"Cartesian axes"), as shown in Fig. 1.2.

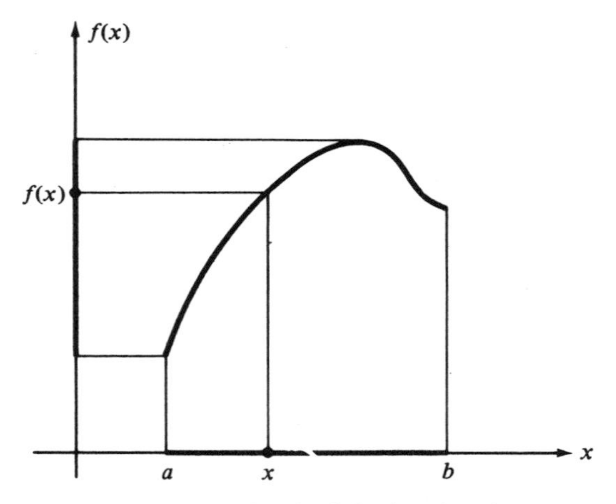

FIGURE 1.2 Graph of the function f.

A *functional*, on the other hand, has as its domain a set of *functions* (and as its range a set of *numbers*); i. e., it is a "function of a function." To illustrate, consider the functional $\mathscr{F}$ defined by

$$\mathscr{F}(u) = \int_0^1 u^2(x)dx$$

where the domain of $\mathscr{F}$ might be defined to be the set of all functions u defined over the interval 0, 1 subject only to the restriction that the integral of u^2 from 0 to 1 does in fact exist. It is not hard to see that the range of $\mathscr{F}$ is the semi-infinite interval $0 \leq \mathscr{F}(u) < \infty$.

[4] It is tempting to emphasize this distinction by calling $\mathscr{L}$ the "total operator" for example, but we choose to simply call it the operator, in agreement with generally accepted nomenclature.

$$\mathbf{B}_1(u) = u(0) - 3u(1) = 0 \tag{1.10b}$$

$$\mathbf{B}_2(u) = u'(0) = 0 \tag{1.10c}$$

If we choose to break up $\phi(x)$ into $\rho(x) + \sigma(x)$, for example, and are able to find functions $v(x)$ and $w(x)$ which satisfy the conditions

$$\mathbf{L}v = \rho; \qquad \mathbf{B}_1(v) = 0, \mathbf{B}_2(v) = 0 \tag{1.11a}$$

$$\mathbf{L}w = \sigma; \qquad \mathbf{B}_1(w) = 0, \mathbf{B}_2(w) = 0 \tag{1.11b}$$

then the solution of (1.10) is given by $u(x) = v(x) + w(x)$. This is easily verified by adding (1.11a) and (1.11b), and using the linearity property of $\mathbf{L}$, $\mathbf{B}_1$, and $\mathbf{B}_2$:

$$\mathbf{L}v + \mathbf{L}w = \rho + \sigma; \qquad \mathbf{B}_j(v) + \mathbf{B}_j(w) = 0 \text{ for } j = 1, 2 \tag{1.12}$$

so that

$$\mathbf{L}(v + w) = \phi; \qquad \mathbf{B}_j(v + w) = 0 \text{ for } j = 1, 2 \tag{1.13}$$

But the conditions (1.13) are exactly those to be satisfied by u, so that $u = v + w$, as claimed.

In the case of the system (1.10), the boundary conditions are **homogeneous** [i. e., all of the c_j's in (1.6) are zero], whereas the differential equation is not (since $\mathbf{L}u$ is equal to ϕ, and not zero). Of course, if both the differential equation and the boundary conditions were homogeneous then the solution would simply be $u = 0$. The presence of a nontrivial solution is, in fact, directly dependent on the presence of some form of non-homogeneity in the problem, either in the differential equation and/or in the boundary conditions.

It is customary, in developing the method of Green's functions, to require that the boundary conditions be homogeneous, with all of the inhomogeneity contained in the differential equation. The case of inhomogeneous boundary conditions is then dealt with by a suitable modification of the basic method. Our development, however, is slightly different, and will not require the homogeneity of the boundary conditions.

Before actually discussing Green's functions it is important to get two more preliminaries out of the way: introduction of (1) the *adjoint operator* and (2) the *delta function*. These will be discussed in the next two sections.

EXERCISES

1.1. Determine whether the following differential operators are linear or nonlinear, by applying the basic definition (1.2):

(a) $\mathbf{L}u = u'' + xu' + 6u$
(b) $\mathbf{L}u = u'' - \sin u$
(c) $\mathbf{L}u = u' + |u|$, where the bars denote "absolute value of" u.

1.2. Determine whether or not the following functionals are linear, by applying the basic definition of linearity:

(a) $\mathscr{F}(u) = u(0) - 3u'(\pi)$ (c) $\mathscr{F}(u) = u^2(0)$

(b) $\mathscr{F}(u) = u(0) + 2$ (d) $\mathscr{F}(u) = \int_0^\pi u(x) \sin x \, dx$

1.3. Prove that if $\mathscr{F}$ is a linear functional (over a domain which contains the function $u = 0$) then $\mathscr{F}(0)$ must be zero.

1.4. What is the range of each of the functionals defined in Exercise 1.2 if their domains consist of the set of all functions, defined over $0 \leq x \leq \pi$, which are differentiable and less than 10 in absolute magnitude?

1.5. What is the range of the function $f(x) = x^2$ if its domain is the interval $-1 \leq x \leq 4$? What is the domain of the function $f(x) = \sin x$ corresponding to the range $0 < f(x) \leq 1$?

2. THE ADJOINT OPERATOR

One way to introduce the so-called **formal adjoint** differential operator $\mathbf{L}^*$, associated with $\mathbf{L}$, is to form the product $v\mathbf{L}u$, and integrate it over the interval of interest.[5] By means of repeated integration by parts, the result can be expressed in the form

$$\int_a^b v\mathbf{L}u \, dx = [\cdots]\Big|_a^b + \int_a^b u\mathbf{L}^*v \, dx \tag{2.1}$$

In this calculation, the functions u and v are understood to be completely arbitrary (except for being sufficiently differentiable for $\mathbf{L}u$ and $\mathbf{L}^*v$ to exist); they are *not* necessarily solutions of any particular differential equation.

To illustrate the procedure, consider the general linear second order differential operator

$$\mathbf{L} = a(x)\frac{d^2}{dx^2} + b(x)\frac{d}{dx} + c(x) \tag{2.2}$$

The product $v\mathbf{L}u$ then consists of three terms. If we integrate the first by parts twice, the second once, and the third not at all, we find easily that

[5] Whereas $\mathbf{L}^*$ is fairly common notation for the formal adjoint of the differential operator $\mathbf{L}$, we note that in some books asterisks are used to denote complex conjugates, with $\mathbf{L}^\dagger$ used for the formal adjoint differential operator.

$$\begin{aligned}\int_a^b v\mathbf{L}u\,dx &= \int_a^b (vau'' + vbu' + vcu)dx \\ &= (vau' + vbu)\Big|_a^b + \int_a^b [-(va)'u' - (vb)'u + vcu]dx \\ &= [vau' + vbu - (va)'u]\Big|_a^b + \int_a^b [(va)''u - (vb)'u + vcu]dx \\ &= [avu' - (av)'u + bvu]\Big|_a^b + \int_a^b u[(av)'' - (bv)' + cv]dx\end{aligned} \tag{2.3}$$

Comparing (2.1) and (2.3), then, we see that

$$\begin{aligned}\mathbf{L}^* v &= (av)'' - (bv)' + cv \\ &= av'' + (2a' - b)v' + (a'' - b' + c)v\end{aligned} \tag{2.4}$$

so that

$$\mathbf{L}^* = a\frac{d^2}{dx^2} + (2a' - b)\frac{d}{dx} + (a'' - b' + c) \tag{2.5}$$

As an aside, we note that the formal adjoint differential operator also arises in another context, namely in connection with an *integrating factor*, say $\rho(x)$, of our equation $\mathbf{L}u = \phi$. First, we say that our nth order differential operator $\mathbf{L}$ is *exact* if there exists an $(n - 1)$th order differential operator $\tilde{\mathbf{L}}$, say, such that $\mathbf{L} = (d/dx)\tilde{\mathbf{L}}$. (The interest in exactness is obvious; if $\mathbf{L}$ is exact, then a first integral of the differential equation is readily available.) If $\mathbf{L}$ is *not* exact then we may consider instead the equation $\rho\mathbf{L}u = \rho\phi$ where $\rho(x)$ is to be chosen so that the new differential operator $\rho\mathbf{L}$ *is* exact. Considering our second order differential operator (2.2), for example, we require that

$$\begin{aligned}\rho au'' + \rho bu' + \rho cu &= (\alpha u' + \beta u)' \\ &= \alpha u'' + (\alpha' + \beta)u' + \beta' u\end{aligned}$$

or,

$$\rho a = \alpha \tag{2.6a}$$

$$\rho b = \alpha' + \beta \tag{2.6b}$$

$$\rho c = \beta' \tag{2.6c}$$

Now, (2.6a) implies that $\alpha' = (\rho a)'$ so that, from (2.6b), $\beta = \rho b - (\rho a)'$. Differentiating this, and using (2.6c), we have $\beta' = (\rho b)' - (\rho a)'' = \rho c$. That is, having eliminated α and β from equations (2.6), we see that integrating factors of $\mathbf{L}$ are found as solutions of the differential equation

$$(a\rho)'' - (b\rho)' + c\rho = 0 \tag{2.7}$$

or, recalling (2.4),

$$\mathbf{L}^* p = 0 \tag{2.8}$$

In the event that $\mathbf{L} = \mathbf{L}^*$ we say that $\mathbf{L}$ is **formally self-adjoint.** For our second order differential operator to be formally self-adjoint we see, by comparing (2.2) and (2.4), that the coefficients must be related such that

$$b = 2a' - b \quad \text{and} \quad c = a'' - b' + c \tag{2.9}$$

These relations will be satisfied if and only if $b = a'$, so that a linear, formally self-adjoint, second order differential operator can be expressed in the form

$$\mathbf{L} = a\frac{d^2}{dx^2} + a'\frac{d}{dx} + c = \frac{d}{dx}\left(a\frac{d}{dx}\right) + c \tag{2.10}$$

Actually, for the case of *second* order differential operators, $\mathbf{L}u = \phi$ can always be thrown into formally self-adjoint form, even if it is not formally self-adjoint to begin with; see Exercise 2.3. However, comparing (2.2) and (2.5) for the case where $a \equiv 0$, observe that *first* order differential operators *cannot* be formally self-adjoint.

For the case where the boundary conditions associated with $\mathbf{L}$ are homogeneous, we define not only a formal adjoint differential operator $\mathbf{L}^*$, but also an **adjoint operator** $\mathscr{L}^*$. Specifically, $\mathscr{L}^*$ is defined by the relation

$$(\mathbf{L}u, v) = (u, \mathbf{L}^* v) \tag{2.11}$$

where the symbol (f, g), generally called the **inner product** (sometimes the *scalar product*, or *dot product*) of f and g, is defined here by

$$(f, g) = \int_a^b f(x)\, g(x)\, dx \tag{2.12}$$

Comparing (2.11) with (2.1), we observe that $\mathscr{L}^*$ consists of the differential operator $\mathbf{L}^*$ plus boundary conditions which are such that the boundary terms, arising through the integration by parts, all vanish.

For example, consider $\mathscr{L}$ to consist of $\mathbf{L} = d/dx$ over the interval $0 \leq x \leq 1$, together with the boundary condition $u(0) = 3u(1)$, that is $\mathbf{B}(u) = u(0) - 3u(1) = 0$. Integrating by parts,

$$\begin{aligned}
(\mathbf{L}u, v) &= \int_0^1 u'v\, dx = (uv)\Big|_0^1 - \int_0^1 uv'\, dx \\
&= u(1)\, v(1) - u(0)\, v(0) + (u, \mathbf{L}^* v) \\
&= u(1)\, [v(1) - 3v(0)] + (u, \mathbf{L}^* v)
\end{aligned} \tag{2.13}$$

Now, since $u(1)$ is not specified, it need not be zero. For the boundary term to vanish, we therefore require of v that $v(1) - 3v(0) = 0$. Thus, $\mathscr{L}^*$ consists of the differential operator $\mathbf{L}^* = -d/dx$ over $0 \leq x \leq 1$, together with the **adjoint boundary condition** $\mathbf{B}^*(v) = 3v(0) - v(1) = 0$.

As one more example, consider $\mathscr{L}$ to consist of $\mathbf{L} = d^2/dx^2$ over the interval $0 \leq x \leq 1$, plus the boundary conditions $\mathbf{B}_1(u) = u(0) = 0$ and $\mathbf{B}_2(u) = u'(1) = 0$. Then

$$(\mathbf{L}u, v) = \int_0^1 u''v\,dx = (u'v - uv')\Big|_0^1 + \int_0^1 uv''\,dx$$
$$= 0 - u(1)v'(1) - u'(0)v(0) + 0 + (u, \mathbf{L}^*v) \tag{2.14}$$

so that $\mathscr{L}^*$ consists of $\mathbf{L}^* = d^2/dx^2$ over $0 \leq x \leq 1$, plus the boundary conditions $\mathbf{B}_1^*(v) = v(0) = 0$ and $\mathbf{B}_2^*(v) = v'(1) = 0$. In this case we have $\mathscr{L}^* = \mathscr{L}$; that is, $\mathbf{L}^* = \mathbf{L}$ *and* $\mathbf{B}_j^* = \mathbf{B}_j$ for $j = 1, 2$. We then say that $\mathscr{L}$ is **self-adjoint.**[6]

EXERCISES

2.1. Determine the adjoint operator $\mathscr{L}^*$ for each of the following operators:

(a) $\mathbf{L} = d^2/dx^2$ over $0 \leq x \leq 1$; $\mathbf{B}_1(u) = u(0) = 0$, $\mathbf{B}_2(u) = u'(0) = 0$.

(b) $\mathbf{L} = d^3/dx^3 - \sin x\, d/dx + 3$ over $0 \leq x \leq \pi$; $\mathbf{B}_1(u) = u(0) = 0$, $\mathbf{B}_2(u) = u'(0) = 0$, $\mathbf{B}_3(u) = u''(0) - 4u(\pi) = 0$.

(c) $\mathbf{L} = d/dx + 1$ over $0 \leq x < \infty$; $\mathbf{B}(u) = u(0) = 0$.

(d) $\mathbf{L} = d^2/dx^2 + 1$ over $0 \leq x \leq 1$; $\mathbf{B}_1(u) = u(0) = 0$, $\mathbf{B}_2(u) = u(1) = 0$.

Which of these operators are self-adjoint? Which of the differential operators are formally self-adjoint?

2.2. If $\mathbf{L}^*$ is the formal adjoint of an nth order differential operator $\mathbf{L}$, show that the formal adjoint of $\mathbf{L}^*$ is, in turn, $\mathbf{L}$; that is, $(\mathbf{L}^*)^* = \mathbf{L}$. (It is under-

[6] The term *Hermitian* is also used in the literature, but perhaps is best reserved for situations where complex-valued functions are present. In that case, a complex-type scalar product

$$(f, g) = \int_a^b \overline{f(x)}g(x)\,dx$$

is often used, where $f(x)$ and $g(x)$ are complex-valued functions of the real variable x, and $\overline{f(x)}$ means the complex conjugate of $f(x)$.

stood, throughout this book, that the letter **L** denotes a *linear* differential operator.)

2.3. Show that even if the differential operator $\mathbf{L} = a(x)(d^2/dx^2) + b(x)(d/dx) + c(x)$ is *not* formally self-adjoint, $\sigma\mathbf{L}$ *will* be if we choose

$$\sigma = \exp\left\{\int \frac{b - a'}{a}\, dx\right\}$$

As an application of this result, show that the inhomogeneous Bessel equation of order k, $x^2u'' + xu' + (x^2 - k^2)u = \phi$, can be made formally self-adjoint by multiplying the whole equation through by $1/x$.

2.4. To see why we required homogeneity of the boundary conditions in defining the adjoint operator $\mathscr{L}^*$, let us try to determine $\mathscr{L}^*$ for a case where the boundary conditions are *not* homogeneous. For example, consider $\mathscr{L}$ to consist of the differential operator $\mathbf{L} = d^2/dx^2 + d/dx + x$ over $0 \leq x \leq 1$, plus the boundary conditions $u(0) = a$, $u(1) = 0$. Show that *three* boundary conditions, $v(0) = v(1) = v'(0) = 0$, are needed for the adjoint operator, whereas $\mathbf{L}^*$ is only of *second* order! This is generally known as a *sticky wicket*.

2.5. Verify that equations (2.9) are satisfied if and only if $b = a'$, as claimed.

2.6. If $\mathbf{L}_1$ and $\mathbf{L}_2$ are linear differential operators, show that $(\mathbf{L}_1 + \mathbf{L}_2)^* = \mathbf{L}_1^* + \mathbf{L}_2^*$, where the combination "$\mathbf{L}_1 + \mathbf{L}_2$" is understood operationally to mean $(\mathbf{L}_1 + \mathbf{L}_2)u = \mathbf{L}_1 u + \mathbf{L}_2 u$.

2.7. It is sometimes convenient to include a *weighting function*, $w(x)$ say, in the definition of the inner product:

$$(f, g) = \int_a^b f(x)\, g(x)\, w(x)\, dx$$

Show that the differential operator $\mathbf{L} = d^2/dx^2 + (1/x)d/dx + 1$, over the interval $a \leq x \leq b$, is formally self-adjoint if the above inner product is used, with $w = x$.

2.8. We proved, in the text, that integrating factors $\rho(x)$ of the differential operator $\mathbf{L} = ad^2/dx^2 + bd/dx + c$, defined over an interval $a \leq x \leq b$, are found as solutions of the equation $\mathbf{L}^*\rho = 0$. A more compact proof, valid for $\mathbf{L}$ of *any* order, is as follows: For all functions $u(x)$ which are sufficiently differentiable over a, b for $\tilde{\mathbf{L}}u$ to exist there, and defined to be zero outside of a, b, we have

$$0 = \tilde{\mathbf{L}}u\Big|_{-\infty}^{\infty} = \int_{-\infty}^{\infty} \left(\frac{d}{dx}\right) \tilde{\mathbf{L}}u\, dx = \int_{-\infty}^{\infty} \rho \mathbf{L}u\, dx = \int_{-\infty}^{\infty} u\mathbf{L}^*\rho\, dx$$

Explain why each step is valid, and why the final result implies that $\mathbf{L}^*\rho = 0$ over a, b.

3. THE DELTA FUNCTION

In physics and engineering we inevitably deal with the notion of "point actions"; i. e., actions which are highly localized in space and/or time. These include point forces and couples in solid mechanics, impulsive forces in rigid body dynamics, point masses in gravitational field theory, point charges and multipoles in electrostatics, and point heat sources and pulses in the theory of heat conduction, to name just a few.

For example, suppose we press a circular coin against the edge of a metal plate of the same thickness, which extends over $y > 0$ and $-\infty < x < \infty$, as shown in Fig. 3.1(a). We press with a unit force, and are interested in the resulting stress field induced in the plate. To set up the boundary value problem we would need to know the force distribution, say $w(x)$ lbs per unit x-length, applied to the edge of the plate. However, the function w, which is probably of the form shown in Fig. 3.1(b), is not known a priori. We do know that it will be quite concentrated, and that

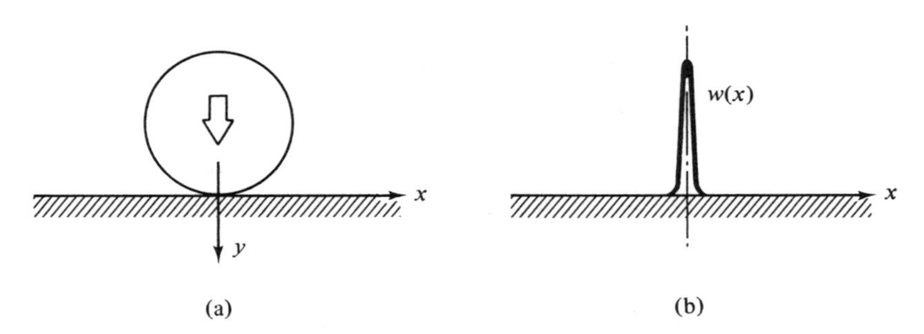

FIGURE 3.1 Force applied to edge of plate.

$$\int_{-\infty}^{\infty} w(x)\, dx = 1 \tag{3.1}$$

so that its net force is unity, but its exact shape must be computed, together with the stress field, as part of the overall problem.

On the other hand, we expect the induced stress field to be quite insensitive to the detailed shape of w,[7] so that we might just as well simplify the problem by assuming, a priori, a definite form for w, such as

[7] More precisely, two highly concentrated force distributions will induce essentially identical stress fields, except in the immediate neighborhood of their point of application, provided that they are ***statically equivalent***; i. e., provided that their resultant forces and couples are identical. (In the present example there will be no couple, by virtue of the symmetry about the point $x = 0$.) This is known as ***Saint-Venant's Principle***.

$$w_k(x) = \begin{cases} \dfrac{k}{2}, & |x| < \dfrac{1}{k} \\ 0, & |x| > \dfrac{1}{k} \end{cases} \tag{3.2}$$

or

$$w_k(x) = \frac{k}{\pi(1 + k^2x^2)} \tag{3.3}$$

for example, where k is a positive number. These are both statically equivalent[8] to the true w, and are quite concentrated for large values of the index k, as shown in Figs. 3.2 and 3.3.

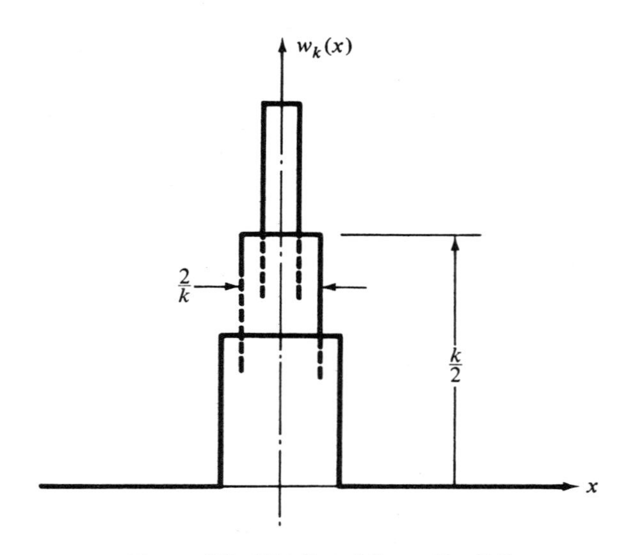

FIGURE 3.2 Distributed force; Eq. (3.2).

Carrying this argument to its logical conclusion, we let $k \to \infty$ and obtain an idealized "point" force of unit strength, say $\delta(x)$, acting at $x = 0$; that is,

$$\delta(x) = \lim_{k \to \infty} w_k(x) \tag{3.4}$$

Mathematically, this definition of the **delta function** is nonsense because the limit is infinite for $x = 0$. Since infinity is not acceptable as a function value, the right-hand side of (3.4) does not define a mathematical function! Nevertheless, the delta function was widely used for many years after being intro-

[8] See Footnote 7. Note that both of these $w_k(x)$'s satisfy (3. 1).

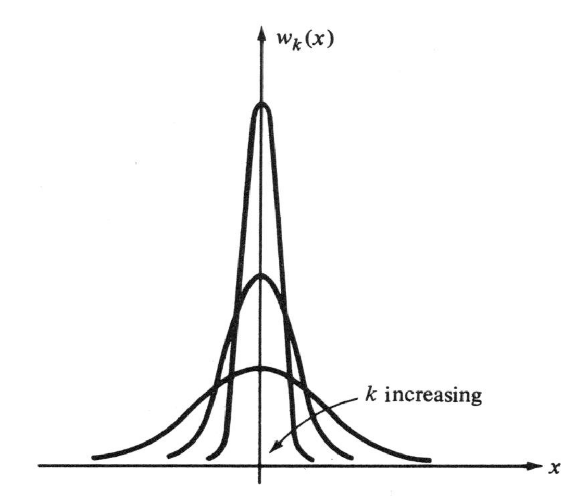

FIGURE 3.3 Distributed force; Eq. (3.3).

duced by P. Dirac,[9] before being defined rigorously as a **generalized function** by L. Schwartz.[10]

We introduce generalized functions through consideration of the functional

$$\int_{-\infty}^{\infty} g(x)\, h(x)\, dx = \mathscr{F}(h) \tag{3.5}$$

That is, to each function h within some prescribed class $\mathscr{D}$ of functions, called the domain of $\mathscr{F}$, the left-hand side assigns a numerical value, $\mathscr{F}(h)$, as discussed in Footnote 3 (pp. 3–4). The choice of the domain $\mathscr{D}$ rests with us. It turns out to be especially fruitful to define $\mathscr{D}$ to consist of functions which are "very smooth," and which decay "very rapidly" at infinity. For definiteness, we will take $\mathscr{D}$ to be the set of all functions, defined over $-\infty < x < \infty$, which are infinitely differentiable, and which vanish outside of some finite interval.

Now, suppose we start specifiying various functionals $\mathscr{F}$. For example, suppose we choose $\mathscr{F}(h)$ to be the integral of h from ξ to ∞. Thus,

$$\int_{-\infty}^{\infty} g(x)\, h(x)\, dx = \int_{\xi}^{\infty} h(x)\, dx \tag{3.6}$$

[9] P.A.M. Dirac, *The Physical Interpretation of the Quantum Mechanics*, Proc. Roy. Soc., London, Section A113, 1926–1927, pp. 621–641.

[10] L. Schwartz, *Théorie des Distributions*, Actualités Scientifiques et Industrielles, Nos. 1091 and 1122, Hermann & Cie, Paris, 1950–1951.

and it follows that the kernel[11] $g(x)$ must be the so-called **Heaviside step function,**[12]

$$H(x - \xi) = \begin{cases} 1, & x > \xi \\ 0, & x < \xi \end{cases} \tag{3.7}$$

shown in Fig. 3.4. In this case, the kernel g turned out to be an "ordinary" function, i. e., in the classical sense.

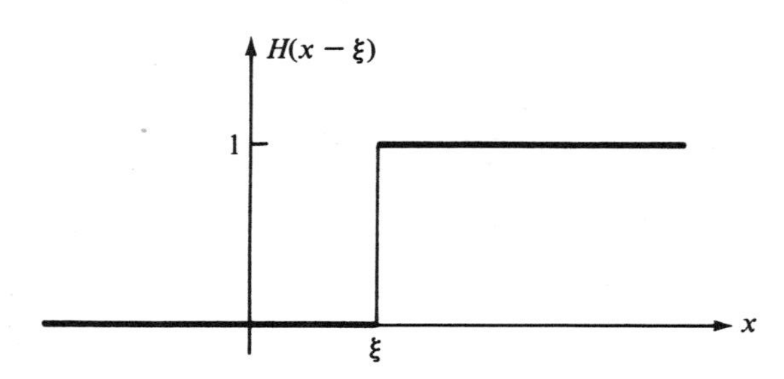

FIGURE 3.4 Heaviside step function.

On the other hand, suppose we choose $\mathscr{F}(h)$ to be $h(0)$, so that

$$\int_{-\infty}^{\infty} g(x)\, h(x)\, dx = h(0) \tag{3.8}$$

In this case it can be shown that no "ordinary" function $g(x)$ exists such that (3.8) holds for all h's in $\mathscr{D}$, and g is therefore understood to be a *generalized function*, defined by (3.8).[13] We emphasize that g is thus defined by its action on a given function h, as spelled out by (3.8) or, in general, by (3.5). That is, we *never talk about the "values" of a generalized function*!

In fact, we shall call the generalized function defined by (3.8) the **delta function.** That is,

[11] Integrands can sometimes be factored into "fixed" and "variable" portions. The fixed part is generally called the *kernel*. In (3.5), we insert various functions h into the integrand, and hence generate the values $\mathscr{F}(h)$. Since the same function g acts on all of these h's, we call it the kernel of the functional $\mathscr{F}$.

[12] The value of $H(x - \xi)$ at $x = \xi$ is of no interest in the present applications; any finite value will do.

[13] Thus, generalized functions arise as the kernels of functionals, more specifically "linear, continuous functionals," called *distributions* by Schwartz. For a precise definition of the term "continuous functional," and a generally deeper discussion, we refer the interested reader to the sources listed in the Suggested Reading. Our object here is twofold: to get at certain concepts and formulas which will be used in our discussion of Green's functions and, at the same time, to give the reader who does wish to pursue this subject a "running start."

$$\int_{-\infty}^{\infty} \delta(x)\, h(x)\, dx = h(0) \tag{3.9}$$

Although rigorous, this definition of $\delta(x)$ (it seems to me) is somewhat abstract, and lacks the homespun flavor of (3.4)—which we recall was unacceptable rigorously because the limit failed to exist for $x = 0$. However, refusing to give up our picturesque w_k sequences, we say that (3.4) is to be understood *symbolically—in the sense that*

$$\lim_{k\to\infty} \int_{-\infty}^{\infty} w_k(x)\, h(x)\, dx = h(0) \tag{3.10}$$

Let us verify this for the w_k sequence (3.2). We have

$$\begin{aligned}\lim_{k\to\infty} \int_{-\infty}^{\infty} w_k(x)\, h(x)\, dx &= \lim_{k\to\infty} \frac{k}{2} \int_{-1/k}^{1/k} h(x)\, dx \\ &= \lim_{k\to\infty} \frac{k}{2} \cdot h(\xi) \cdot \frac{2}{k} = h(0)\end{aligned} \tag{3.11}$$

where, since the allowable h's are continuous, the next-to-last step follows from the mean value theorem of the integral calculus, with ξ some number between $-1/k$ and $+1/k$; clearly, ξ is forced to zero as $k \to \infty$. We therefore say that $w_k(x)$ defined by (3.2) is a *δ-sequence.*

Observe that we take the limit of the integral, in (3.10), *not* the integral of the limit. Nowhere in (3.11) are we confronted with lim $w_k(x)$, so that the mathematical shortcoming of (3.4) is completely sidestepped.

Verification of (3.10) for the w_k sequence (3.3) is not as simple as it was for the piecewise-constant w_k's of (3.2). Instead of pursuing this, let us prove a more general result, which will apply to a wide class of δ-sequences.

Assertion. *If $w(x)$ is a nonnegative function satisfying $\int_{-\infty}^{\infty} w(x)\, dx =$ 1, then $kw(kx) \equiv w_k(x)$ is a δ-sequence.*

Before proving this *assertion* let us use it to verify the sequence (3.3). Noting that

$$w(x) = \frac{1}{\pi(1 + x^2)} \tag{3.12}$$

is nonnegative and has unit area, it follows that

$$w_k(x) = kw(kx) = \frac{k}{\pi(1 + k^2x^2)} \tag{3.13}$$

is, in fact, a δ-sequence. Now let us prove the *assertion.*

Proof. First, we show that lim $w_k(x) = 0$ as $k \to \infty$, for each fixed $x \neq 0$. Since w is nonnegative, and its integral from $-\infty$ to $+\infty$ exists, it follows that $w(x)$ must be $O(1/x^{1+\alpha})$ as $|x| \to \infty$, where $\alpha > 0$.[14] Thus $w_k(x) = kw(kx) = O(k \cdot k^{-1-\alpha}x^{-1-\alpha}) = O(k^{-\alpha}) \to 0$ as $k \to \infty$ for fixed $x \neq 0$.

Now let us express

$$\begin{aligned}\lim_{k\to\infty} \int_{-\infty}^{\infty} w_k(x)\, h(x)\, dx &= \lim_{k\to\infty} \int_{-\infty}^{\infty} w_k(x)\, [h(x) - h(0)]\, dx \\ &\quad + \lim_{k\to\infty} \int_{-\infty}^{\infty} w_k(x)\, h(0)\, dx. \\ &= I + J, \text{ say.}\end{aligned} \tag{3.14}$$

First consider J.

$$\begin{aligned}J &= h(0) \lim_{k\to\infty} \int_{-\infty}^{\infty} kw(kx)\, dx \\ &= h(0) \lim_{k\to\infty} \int_{-\infty}^{\infty} w(\xi)\, d\xi = h(0) \lim_{k\to\infty} 1 = h(0)\end{aligned} \tag{3.15}$$

We now hope to show that $I = 0$ so that the right-hand side of (3.14) will be $h(0)$, as desired. Choose any number $\beta > 0$. Since h is assumed to be continuous at $x = 0$, there must exist (by the definition of continuity) a number $\gamma > 0$ such that $|h(x) - h(0)| < \beta$ whenever $|x - 0| = |x| < \gamma$. Breaking up the I integral,

$$I = \lim_{k\to\infty} \int_{-\infty}^{-\gamma} + \lim_{k\to\infty} \int_{-\gamma}^{\gamma} + \lim_{k\to\infty} \int_{\gamma}^{\infty} \equiv I_1 + I_2 + I_3 \tag{3.16}$$

we observe that $I_1 = I_3 = 0$ since $w_k(x) \to 0$ uniformly (see Exercise 3.5), over $-\infty < x \leq -\gamma$ and $\gamma \leq x < \infty$, as $k \to \infty$. Estimating I_2, we have

$$\begin{aligned}|I_2| &\leq \beta \lim_{k\to\infty} \int_{-\gamma}^{\gamma} w_k(x)\, dx = \beta \lim_{k\to\infty} \int_{-\gamma}^{\gamma} kw(kx)\, dx \\ &= \beta \lim_{k\to\infty} \int_{-k\gamma}^{k\gamma} w(\xi)\, d\xi = \beta\end{aligned} \tag{3.17}$$

Finally, since β can be chosen as small as we like, it follows that $I_2 = 0$, yielding the desired result. This completes the proof.

We point out that whereas $\delta(x)$ acts at $x = 0$, we can have it act at any desired point simply by shifting the argument. Thus, $\delta(x - \xi)$ acts at $x = \xi$,

$$\int_{-\infty}^{\infty} \delta(x - \xi)\, h(x)\, dx = h(\xi) \tag{3.18}$$

[14] The "big oh" is the Bachmann-Landau *order of magnitude* symbol. By $f(x) = O[g(x)]$ as $x \to x_0$, say, we mean that $f(x)/g(x)$ is bounded as $x \to x_0$. As a simple example, we have $\sin 6x = O(x)$ as $x \to 0$, and $O(1)$ as $x \to \infty$. According to *asymptotic* notation, by contrast, $f(x) \sim g(x)$ as $x \to x_0$ means that $[f(x)/g(x)] \to 1$ as $x \to x_0$. Thus, $\sin 6x \sim 6x$ as $x \to 0$; for the limit $x \to \infty$, the asymptotic notation is of no help since there is no "simpler" function to which $\sin 6x$ is asymptotic.

As a matter of fact, it is even permissible, within the framework of generalized function theory, to "differentiate" the delta function. Recalling the basic definition (3.5) of a generalized function g, we see that defining the "derivative g'" really amounts to deciding what functional $\mathscr{F}(h)$ is defined by

$$\int_{-\infty}^{\infty} g'(x)\, h(x)\, dx \tag{3.19}$$

Staying as close as possible to "ordinary" function theory, so that our result will still be valid in the event that g is an "ordinary" function, let us "integrate by parts":

$$\int_{-\infty}^{\infty} g'(x)\, h(x)\, dx = g(x)\, h(x)\Big|_{-\infty}^{\infty} - \int_{-\infty}^{\infty} g(x)\, h'(x)\, dx \tag{3.20}$$

Note that we cannot *prove* that the integration by parts formula of classical function theory is valid for generalized functions since the formula contains g', which has not yet even been defined. Indeed, we are using (3.20) to define g'!

Let us examine the two terms on the right-hand side. The integral term is very nice because we already know how g acts on functions.[15] The boundary term, however, causes some pain because it involves knowing the *values* of g, which are not defined. We circumvent this difficulty by simply throwing the boundary term away, and *defining* g' by the formula

$$\int_{-\infty}^{\infty} g'(x)\, h(x)\, dx = -\int_{-\infty}^{\infty} g(x)\, h'(x)\, dx \tag{3.21}$$

This is, in fact, valid in the event that g is an "ordinary" function, since then the boundary term drops out by virtue of the fact that the allowable h's all vanish outside some finite interval.

To illustrate (3.21), let us take $g(x) = \delta(x - \xi)$. Then $\delta'(x - \xi)$ is defined by

$$\begin{aligned} \int_{-\infty}^{\infty} \delta'(x - \xi)\, h(x)\, dx &= -\int_{-\infty}^{\infty} \delta(x - \xi)\, h'(x)\, dx \\ &= -h'(\xi) \end{aligned} \tag{3.22}$$

Repeating the process, the "jth derivative $\delta^{(j)}(x - \xi)$" is defined by the formula

$$\int_{-\infty}^{\infty} \delta^{(j)}(x - \xi)\, h(x)\, dx = (-1)^j h^{(j)}(\xi) \tag{3.23}$$

[15] More precisely, we know how g acts on functions in the set $\mathscr{D}$. Because of the way we have defined $\mathscr{D}$, the fact that h is in $\mathscr{D}$ implies that h' is also. Thus we do know how g acts on h'.

It is interesting to take $g(x) = H(x - \xi)$ in (3.21). Whereas the "ordinary" derivative $H'(x - \xi)$ does not exist, because of the discontinuity in $H(x - \xi)$ at $x = \xi$, $H'(x - \xi)$ does exist as a generalized function. Namely,

$$\begin{aligned} \int_{-\infty}^{\infty} H'(x - \xi)\, h(x)\, dx &= -\int_{-\infty}^{\infty} H(x - \xi)\, h'(x)\, dx \\ &= -\int_{\xi}^{\infty} h'(x)\, dx = h(\xi) \end{aligned} \tag{3.24}$$

since $h(\infty) = 0$. But, recalling that

$$\int_{-\infty}^{\infty} \delta(x - \xi)\, h(x)\, dx = h(\xi) \tag{3.25}$$

we see that

$$H'(x - \xi) = \delta(x - \xi) \tag{3.26}$$

Equations like (3.26), which involve generalized functions, are understood in the sense that if we multiply through by an arbitrary h in $\mathscr{D}$, integrate from $-\infty$ to $+\infty$, and use the functional behavior of the generalized functions to evaluate the integrals, then the result will be an equality. As another illustration of this convention, observe that

$$x\delta(x) = 0 \tag{3.27}$$

Since $xh(x)$ is in $\mathscr{D}$, (3.27) follows from the fact that

$$\int_{-\infty}^{\infty} x\delta(x)\, h(x)\, dx = [xh(x)]|_{x=0} = 0 \tag{3.28}$$

Now let us proceed with our development of the method of Green's functions.

EXERCISES

3.1. Sketch the following sequences, and verify that they are δ-sequences:

(a) $w_k(x) = \begin{cases} k, & 0 < x < 1/k \\ 0, & x \leq 0 \text{ and } x \geq 1/k \end{cases}$

(b) $w_k(x) = \begin{cases} 0, & |x| > 1/2k \\ 4k^2 x + 2k, & -1/2k \leq x \leq 0 \\ -4k^2 x + 2k, & 0 \leq x \leq 1/2k \end{cases}$

(c) $w_k(x) = ke^{-k^2x^2}/\sqrt{\pi}$. *Hint:* $\int_{-\infty}^{\infty} e^{-x^2}\, dx = \sqrt{\pi}$

(d) $w_k(x) = \begin{cases} -k, & |x| < 1/2k \\ 2k, & 1/2k \le |x| \le 1/k \\ 0, & |x| > 1/k \end{cases}$

Hint: Our *assertion* does not apply in this case.
Instead, follow the lines of equations (3.11).

3.2. Inverting the order of integration in the Fourier integral theorem

$$f(\xi) = \frac{1}{2\pi}\int_{-\infty}^{\infty} e^{i\alpha\xi}\,d\alpha \int_{-\infty}^{\infty} f(x)e^{-i\alpha x}\,dx$$

and noting that the formula is valid for all f's in our set $\mathscr{D}$ (called the set of *testing functions* in the literature), obtain the following integral form for the delta function.

$$\delta(x - \xi) = \frac{1}{2\pi}\int_{-\infty}^{\infty} e^{i(\xi - x)\alpha}\,d\alpha$$

Expressing

$$\lim_{k\to\infty} \frac{1}{2\pi}\int_{-k}^{k} e^{i(\xi - x)\alpha}\,d\alpha = \lim_{k\to\infty} \frac{\sin k(x - \xi)}{\pi(x - \xi)}$$

suggests that

$$w_k(x) = \frac{\sin k(x - \xi)}{\pi(x - \xi)}$$

is a δ-sequence. Sketch $w_k(x)$.

3.3. Sketch the function

$$K(x - \xi; \mu) = \frac{1}{2\pi}\frac{1 - \mu^2}{1 + \mu^2 - 2\mu\cos(x - \xi)}, \quad 0 \le \mu < 1$$

over $0 \le x \le 2\pi$, for some fixed value of ξ in the interval, and a few values of the parameter μ; for example, $\mu = 0, 0.5$, and 0.99. Noting that

$$\int_0^{2\pi} \frac{dx}{1 - a\cos x} = \frac{2\pi}{\sqrt{1 - a^2}}$$

if $a^2 < 1$, verify the symbolic equation

$$\lim_{\mu\to 1} K(x - \xi; \mu) = \delta(x - \xi)$$

That is,

$$\lim_{\mu\to 1}\int_0^{2\pi} K(x - \xi; \mu)\,h(x)\,dx = h(\xi)$$

3.4. Following the lines of equations (3.11), verify that

$$w_k(x) = \begin{cases} -k^2, & 0 \leq x < 1/k \\ k^2, & -1/k < x < 0 \\ 0, & |x| \geq 1/k \end{cases}$$

constitutes a δ'-sequence. Show that $\delta'(x)$ may be interpreted physically, in terms of force distributions for example, as a point *couple* of unit strength acting at $x = 0$ [whereas $\delta(x)$ is a point *force* of unit strength acting at $x = 0$]. In electrostatics we would call it a *dipole*.

3.5. In reference to equation (3.10), it should be carefully understood that the position of the limit *outside* the integral sign is important. In other words, the limit of an integral is not necessarily equal to the integral of the limit. Sufficient conditions for the equality

$$\lim_{k\to\infty} \int_a^b f_k(x)\, dx = \int_a^b \lim_{k\to\infty} f_k(x)\, dx \tag{A}$$

are that $\lim f_k(x)$ converge to $F(x)$, say, **uniformly** over $a \leq x \leq b$. [That is, to each positive ϵ, no matter how small, there exists an N such that $F(x) - \epsilon < f_k(x) < F(x) + \epsilon$ over $a \leq x \leq b$, for all k's exceeding N.] To illustrate, consider the simple case where $a = 0$, $b = 1$ and f_k is as shown in the figure.

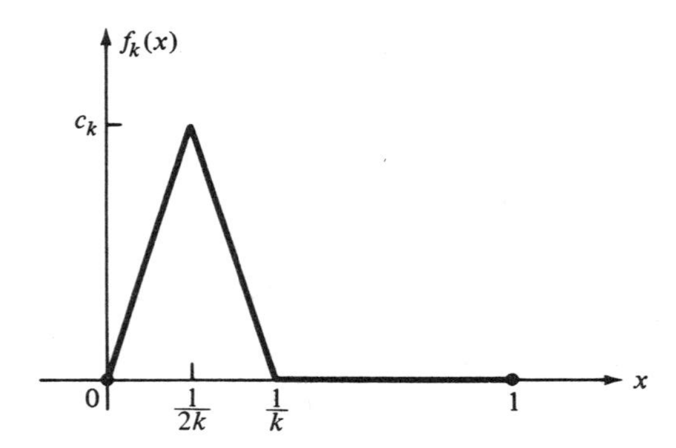

(a) Choosing $c_k = 2k$, show that the f_k sequence does converge but *not* uniformly, and that the left side of (A) is unity whereas the right side is zero. [*Hint:* The limit function $F(x)$ is identically zero over $0 \leq x \leq 1$, despite the fact that the triangular "spike" grows with k! To understand this, observe that for each *fixed* $x > 0$, no matter how small, k eventually reaches a value such that the spike is completely to the left of x, and remains there for all subsequent k's. For $x = 0$, each f_k is zero. Thus $\lim_{k\to\infty} f_k(x) = 0$ for $0 \leq x \leq 1$.]

(b) Choosing $c_k = 1$ for all k's, show that the f_k's do not converge uniformly, and that both sides of (A) are zero. [This shows that uniform convergence of the f_k-sequence is not a *necessary* condition for the validity of (A).]

(c) Choosing $c_k = 1/k$, show that the f_k's do converge uniformly, and that both sides of (A) are zero,

3.6. Prove that

(a) $e^x \delta(x) = \delta(x)$

(b) $x\delta'(x) = -\delta(x)$

(c) $(d^4/dx^4)|x|^3 = 12\delta(x)$

Hint: $|x| = 2xH(x) - x$.

3.7. Using Leibnitz's formula (see Exercise 4.2), show that

$$u(x) = \int_0^1 |x - \xi|\, \phi(\xi)\, d\xi$$

satisfies the differential equation $u''(x) = 2\phi(x)$.

3.8. Derive the formula

$$\int H(x - \xi)\, dx = (x - \xi)H(x - \xi) + \text{constant}$$

3.9. Show that $\delta(\xi - x) = \delta(x - \xi)$. More generally, show that if $f(\xi)$ is a monotonic increasing or decreasing function of ξ which vanishes for $\xi = x$, then $\delta[f(\xi)] = \delta(\xi - x)/|f'(x)|$. [By monotone increasing, for example, we mean that $f(\xi_2) > f(\xi_1)$ whenever $\xi_2 > \xi_1$.] *Hint:* For a start, set $f(\xi)$ equal to a new variable, say u.

3.10. Find the second derivative of $f(x)$, in terms of generalized functions, where $f(x)$ is defined to equal x over $|x| \leq 1$, $+1$ for $x > 1$ and -1 for $x < -1$.

4. THE GREEN'S FUNCTION METHOD

We now turn to the development of the Green's function method for the solution of nth order linear ordinary differential equations, with boundary conditions which consist of linear combinations of the unknown and its derivatives, through order $n - 1$.

We have tried to keep these examples as simple as possible, while at the same time exposing the main ideas associated with the method. Thus, whereas we sometimes appear to be "cracking peanuts with a sledgehammer," we believe that an understanding of these examples will leave the reader in a position to deal with more complicated applications.

Example 1. *Loaded String.* Consider, first, the simple boundary value problem

$$u''(x) = \phi(x); \qquad u(0) = u(1) = 0 \tag{4.1}$$

where $\phi(x)$ is prescribed. When given a problem in a purely mathematical form, it is often helpful to first recast it in physical terms so that we can benefit from whatever physical insight may be at our disposal. Accordingly, we note (see Exercise 4.1) that $u(x)$ in (4.1) may be regarded as the static deflection of a string, stretched under unit tension between fixed endpoints, and subjected to a force distribution $\phi(x)$ pounds per unit length, as sketched in Fig. 4.1. This viewpoint will permit a simple interpretation of the resulting Green's function.

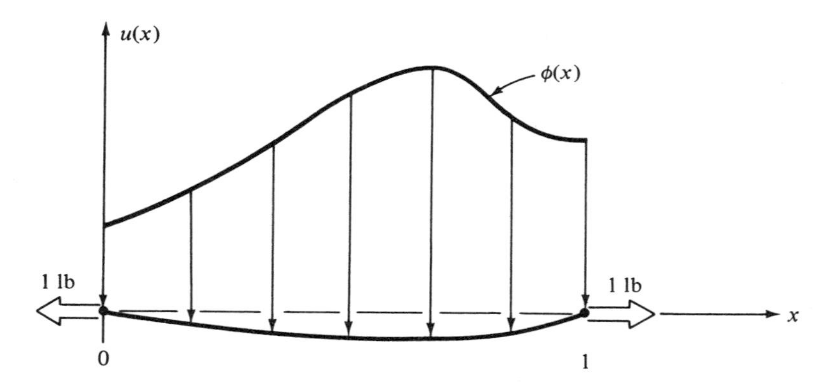

FIGURE 4.1 Loaded string interpretation of Eq. (4.1).

Our starting point, in the solution of (4.1) by the method of Green's functions, is equation (2.1) with "v" replaced by G, the so-called "Green's function." Replacing the integration variable x by a dummy variable ξ, we have

$$\int_0^1 G\mathbf{L}u \, d\xi = [\cdots]\Big|_0^1 + \int_0^1 u\mathbf{L}^* G \, d\xi \tag{4.2}$$

The boundary terms and $\mathbf{L}^*$ are found easily by actually carrying out the integration-by-parts of the left-hand side, with $\mathbf{L} = d^2/d\xi^2$. At the same time, $\mathbf{L}u$ in the left-hand side is equal to ϕ, so that we have

$$\int_0^1 G\phi \, d\xi = (Gu' - uG_\xi)\Big|_0^1 + \int_0^1 uG_{\xi\xi} \, d\xi \tag{4.3}$$

where we have used subscripts on G instead of primes to denote differentiation since, as we shall see, G will be a function not only of the dummy variable ξ, but also of the fixed quantity x. Writing out the boundary terms in (4.3),

$$\begin{aligned}\int_0^1 G(\xi, x)\, \phi(\xi) \, d\xi = G(1, x)\, u'(1) - u(1)\, G_\xi(1, x) - G(0, x)\, u'(0) \\ + u(0)\, G_\xi(0, x) + \int_0^1 u(\xi)\, \mathbf{L}^* G \, d\xi\end{aligned} \tag{4.4}$$

Now, if we choose G cleverly, (4.4) can provide us with the solution to our original problem (4.1). Specifically, if we require that $\mathbf{L}^* G = \delta(\xi - x)$ then the last term is simply $u(x)$. Now look at the four boundary terms. The first and third are somewhat unwelcome since, while $u(0)$ and $u(1)$ are given, $u'(0)$ and $u'(1)$ are not known a priori. However, we can remove these two terms by requiring that $G(\xi, x)$ satisfy the boundary conditions $G(0, x) = G(1, x) = 0$. Thus, if our Green's function satisfies the boundary value problem

$$\mathbf{L}^* G = G_{\xi\xi}(\xi, x) = \delta(\xi - x) \tag{4.5a}$$

$$G(0, x) = G(1, x) = 0 \tag{4.5b}$$

then the solution of (4.1) is given by

$$u(x) = \int_0^1 G(\xi, x)\, \phi(\xi) \, d\xi \tag{4.6}$$

Finally, it remains to calculate the Green's function. Integrating (4.5a) [with the help of (3.26) and Exercise 3.8], with x regarded as fixed,

$$\begin{aligned} G_\xi &= H(\xi - x) + A \\ G &= (\xi - x)\, H(\xi - x) + A\xi + B \end{aligned} \tag{4.7}$$

Imposing the boundary conditions (4.5b),

$$G(0, x) = 0 = 0 + 0 + B \tag{4.8}$$

$$G(1, x) = 0 = (1 - x) + A + B \tag{4.9}$$

so that $B = 0$, $A = x - 1$, and hence

$$G(\xi, x) = (\xi - x)\, H(\xi - x) + (x - 1)\xi \tag{4.10}$$

COMMENT 1. Having recast the statement (4.1) in terms of a loaded string, the Green's function governed by (4.5) admits an obvious physical interpretation. Specifically, $G(\xi, x)$ is the deflection, as a function of ξ, due *not* to the load distribution ϕ but rather due to a point load of unit strength, $\delta(\xi - x)$, acting at the point $\xi = x$. Plotting (4.10) in Fig. 4.2, the result is, in fact, in accord with what we would expect on physical grounds.

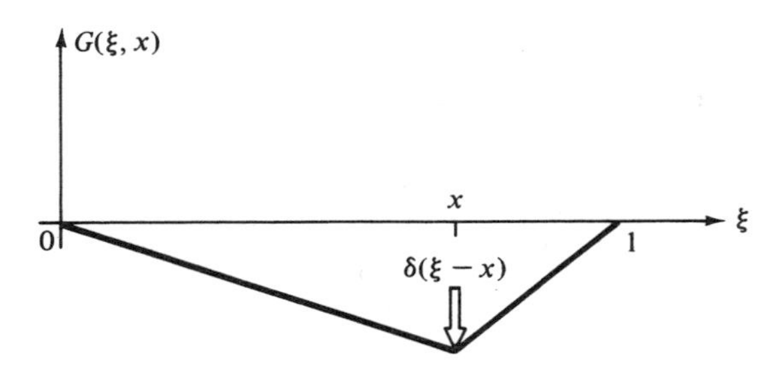

FIGURE 4.2 Green's function for Example 1.

Rewriting (4.10) in the form

$$G(\xi, x) = \begin{cases} (x - 1)\xi, & \xi \leq x \\ (\xi - 1)x, & \xi \geq x \end{cases} \tag{4.11}$$

we observe that G is **symmetric**; $G(\xi, x) = G(x, \xi)$. That is, the deflection at ξ due to a unit load at x is equal to the deflection at x due to a unit load at ξ. This striking result, often referred to as "Maxwell reciprocity," is illustrated graphically in Fig. 4.3. (We will examine this point in more detail in Comment 3.) We are now in a position to interpret (4.6). Because of the above reciprocity, $G(\xi, x)$ is the deflection, as a function of x, due to a unit load at ξ. Clearly, then, $G(\xi, x)\,\phi(\xi)\,d\xi$ is the deflection due to an incremental load $\phi(\xi)\,d\xi$ at ξ, so that (4.6) represents the **superposition** of the resulting incremental deflections.

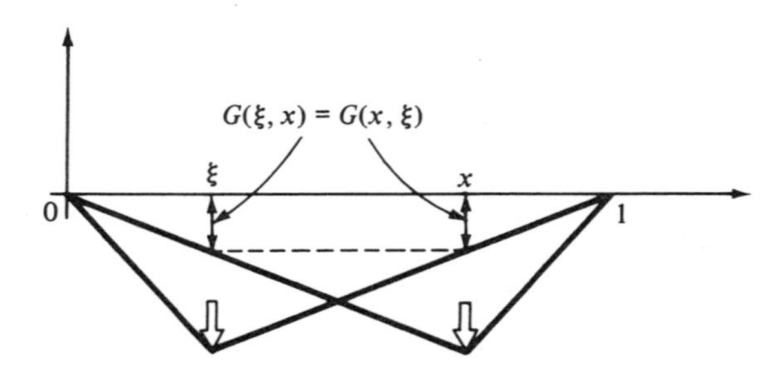

FIGURE 4.3 Maxwell reciprocity.

As a result, G is often called the *influence function.* It should be carefully noted that the superposition nature of (4.6) is a consequence of the linearity of the operator $\mathscr{L}$ of the system (4.1); namely, $\mathbf{L} = d^2/dx^2$, $\mathbf{B}_1(u) = u(0)$, and $\mathbf{B}_2(u) = u(1)$. In fact, it might be a good idea to reread the discussion of superposition in Section 1, starting with the paragraph following equation (1.8).

COMMENT 2. We point out that the boundary terms do not always vanish. For example, suppose we modify our problem (4.1) by changing the right end condition to $u(1) = \alpha$, say. Applying the same reasoning as before, we again want G to satisfy (4.5), so that the Green's function is the same as before. The only difference is that the second boundary term in (4.4) does not vanish. Specifically,

$$u(1)\, G_\xi(1, x) = \alpha G_\xi(\xi, x)|_{\xi=1} = \alpha[1 + (x - 1)]|_{\xi=1} = \alpha x \tag{4.12}$$

so that instead of (4.6), we have

$$u(x) = \alpha x + \int_0^1 G(\xi, x)\, \phi(\xi)\, d\xi \tag{4.13}$$

with G given, again, by (4.10). Observe that our basic procedure remained unchanged here, even though the boundary conditions were no longer homogeneous.

COMMENT 3. Let us examine the observed reciprocity of G in more generality. Recall that G satisfies a system of the form

$$\mathbf{L}^* G(\xi, x) = \delta(\xi - x), \text{ plus suitable homogeneous boundary conditions, say B. C.} \tag{4.14}$$

Denoting the operator of (4.14) (i.e., the differential operator $\mathbf{L}^*$ plus the homogeneous boundary conditions B. C.) as $\mathscr{G}$, we note that associated with $\mathscr{G}$ is its adjoint $\mathscr{G}^*$, consisting of the differential operator $(\mathbf{L}^*)^* = \mathbf{L}$ (recall Exercise 2.2) plus certain other homogeneous boundary conditions, say B. C.*, as required by (2.11). Thus, corresponding to $\mathscr{G}^*$ there is a so-called adjoint Green's function, say G^*, which satisfies

$$(\mathbf{L}^*)^* G^*(\xi, x_0) = \mathbf{L} G^*(\xi, x_0) = \delta(\xi - x_0), \tag{4.15}$$
$$\text{plus suitable homogeneous boundary conditions, B. C.*}$$

where we have introduced the zero subscript on x since the x_0's in (4.15) need not be the same as the x's in (4.14).

Now, taking u and v in (2.11) to be $G^*(\xi, x_0)$ and $G(\xi, x)$ respectively, and using (2.12), (4.14), and (4.15) we have

$$\int_a^b \delta(\xi - x_0)\, G(\xi, x)\, d\xi = \int_a^b G^*(\xi, x_0)\, \delta(\xi - x)\, d\xi$$

or

$$G(x_0, x) = G^*(x, x_0) \tag{4.16}$$

In our Example 1, $\mathscr{G}$ was self-adjoint, so that (4.16) reduced to the simple symmetry condition

$$G(x_0, x) = G(x, x_0) \tag{4.17}$$

EXERCISES

4.1. Verify the equivalence between our differential equation $u'' = \phi$ and the loaded string interpretation of Fig. 4.1. That is, consider the vertical equilibrium of elements of the string, subject to the linearizing assumption of small slopes, i. e., $|u'(x)| \ll 1$ over $0 \leq x \leq 1$, where the double inequality means "very much smaller than."

4.2. Verify, by direct differentiation, that

$$u(x) = \int_0^1 G\phi\, d\xi = (x - 1) \int_0^x \xi\phi(\xi)\, d\xi + x \int_x^1 (\xi - 1)\, \phi(\xi)\, d\xi$$

does in fact satisfy the boundary value problem (4.1). [Recall the **Leibnitz formula** for differentiation of the integral

$$I(x) = \int_{a(x)}^{b(x)} f(\xi, x)\, d\xi$$

with respect to x, namely,

$$\frac{dI}{dx} = \int_{a(x)}^{b(x)} \frac{\partial f}{\partial x}\, d\xi + f[b(x), x]\frac{db}{dx} - f[a(x), x]\frac{da}{dx}$$

provided that f and $\partial f/\partial x$ are continuous functions of both arguments, and that $a(x)$ and $b(x)$ are differentiable. As a special case, a and/or b can of course be constants; *finite* constants, that is, since the case of infinite limits is somewhat singular and requires further restrictions on the integrand. In the event that $a(x)$ is a finite constant, $b(x) = x$, and f depends only on ξ, the Leibnitz formula reduces to the well-known **fundamental theorem of the integral calculus**

$$\frac{d}{dx}\int_a^x f(\xi)\,d\xi = f(x)$$

provided that f is continuous.]

4.3. Instead of using the method of Green's functions, solve (4.1) by direct integration. That is, integrating twice and then invoking the required end conditions yields

$$u(x) = \int_0^x \int_0^\xi \phi(\tau)\,d\tau\,d\xi - x\int_0^1 \int_0^\xi \phi(\tau)d\tau\,d\xi$$

Inverting the order of integration in both these integrals and shaking the result slightly lead to the form quoted in Exercise 4.2. Alternatively, the two integrals above can be integrated by parts; in the first integral, for example, set

$$\int_0^\xi \phi(\tau)\,d\tau \equiv \text{"}u\text{"} \quad \text{and} \quad d\xi \equiv \text{"}dv\text{"}$$

[It's interesting to note that the direct integration of (4.1) is not as trivial as we might expect. The difficulty is due, in fact, to the "boundary value" nature of the problem, i. e., the presence of conditions at *both* endpoints. By contrast, if we replace the boundary conditions by "initial" conditions, say $u(0) = a$ and $u'(0) = b$, then direct integration *is* trivial.]

Example 2. *A More Complicated Operator*. As our second example, we have chosen the problem

$$u'' + 3u' + 2u = \phi; \qquad \begin{aligned} u(1) &= 2u(0) \\ u'(1) &= a \end{aligned} \tag{4.18}$$

because it illustrates several important features not present in Example 1. Specifically, **L** is *not* formally self-adjoint, the first boundary condition is *mixed* (i. e., it involves values at *both* endpoints), and the differential equation for the Green's function is not directly integrable, as it was in Example 1. Of these features, the first two will require no modification at all in the method outlined in Example 1.

Proceeding as before, we integrate by parts:

$$\begin{aligned}
\int_0^1 G\mathbf{L}u\,d\xi &= \int_0^1 G(u'' + 3u' + 2u)\,d\xi \\
&= (Gu' - G_\xi u + 3Gu)\Big|_0^1 + \int_0^1 u(G_{\xi\xi} - 3G_\xi + 2G)\,d\xi \\
&= u(0)\,[6G(1, x) - 3G(0, x) - 2G_\xi(1, x) + G_\xi(0, x)] \\
&\quad + aG(1, x) - G(0, x)\,u'(0) + \int_0^1 u\mathbf{L}^*G\,d\xi
\end{aligned} \tag{4.19}$$

We next require that G satisfy $\mathbf{L}^* G = \delta(\xi - x)$, plus boundary conditions that are chosen so as to remove the unwelcome $u(0)$ and $u'(0)$ terms on the right-hand side of (4.19). Thus, if our Green's function satisfies

$$\mathbf{L}^* G = G_{\xi\xi} - 3G_\xi + 2G = \delta(\xi - x) \tag{4.20}$$

together with the boundary conditions

$$G(0, x) = 0 \tag{4.21}$$

$$6G(1, x) - 2G_\xi(1, x) + G_\xi(0, x) = 0 \tag{4.22}$$

then the solution from (4.19), with $\mathbf{L}u$ replaced by ϕ will be given by

$$u(x) = -aG(1, x) + \int_0^1 G(\xi, x)\, \phi(\xi)\, d\xi \tag{4.23}$$

Let us now determine G. We see that equation (4.20) is not quite as simple as the differential equation (4.5), which we were able to solve by direct integration. Recalling that $\delta(\xi - x) = 0$ for all $\xi \neq x$, it will be convenient to split the interval into two parts, $0 \leq \xi < x$ and $x < \xi \leq 1$ (in each of which $\mathbf{L}^* G = 0$).[16] Clearly then,

$$G(\xi, x) = \begin{cases} Ae^\xi + Be^{2\xi}, & 0 \leq \xi < x & \text{(4.24a)} \\ Ce^\xi + De^{2\xi}, & x < \xi \leq 1 & \text{(4.24b)} \end{cases}$$

To determine the coefficients A, B, C, and D, we have the two boundary conditions which become

$$A + B = 0 \tag{4.25}$$

$$A + 2B + 4eC + 2e^2 D = 0 \tag{4.26}$$

plus two "matching" conditions, which blend the two parts of (4.24) suitably at $\xi = x$. Suitable matching conditions will become apparent if we integrate the differential equation (4.20) from "$x - 0$" to "$x + 0$":[17]

$$\begin{aligned} \int_{x-0}^{x+0} (G_{\xi\xi} - 3G_\xi + 2G)\, d\xi &= \int_{x-0}^{x+0} \delta(\xi - x)\, d\xi \\ G_\xi \Big|_{x-0}^{x+0} - 3G \Big|_{x-0}^{x+0} + 2\int_{x-0}^{x+0} G d\xi &= 1 \end{aligned} \tag{4.27}$$

If we demand that

[16] This step, as we have stated it, is somewhat heuristic since it involves talking about the *values* of $\delta(\xi - x)$. For rigorous treatment of this point we refer the reader to B. Friedman, *Principles and Techniques of Applied Mathematics*, Wiley & Sons, New York, 1956, Chapter 3.

[17] By $f(x - 0)$, for example, we mean the "left-handed" limit of f at x, i. e., the limit of $f(x - \epsilon)$ as $\epsilon \to 0$ through positive values.

$$G(\xi, x) \text{ be a continuous function of } \xi \text{ at } \xi = x \tag{4.28}$$

then the second and third terms of (4.27) drop out, exposing the jump condition on the slope,

$$G_\xi \Big|_{x-0}^{x+0} = 1 \tag{4.29}$$

That is, G is continuous at $\xi = x$, but has a "kink" there. This could also have been seen heuristically from (4.20) as follows. The presence of the delta function on the right-hand side implies that the solution G must be somewhat singular at $\xi = x$. Since differentiation makes singularities even more singular, we expect the delta function behavior to be borne by the highest derivative in the left-hand side, namely $G_{\xi\xi}$. Working backward then, G_ξ should have a finite discontinuity at $\xi = x$, and G should be continuous there—but with a "kink."[18]

In terms of A, B, C, D, conditions (4.28) and (4.29)[19] become

$$A + Be^x - C - De^x = 0 \tag{4.30}$$

$$-Ae^x - 2Be^{2x} + Ce^x + 2De^{2x} = 1 \tag{4.31}$$

(Remember that x is regarded as a constant here. The running variable is ξ.) Solving equations (4.25), (4.26), (4.30), and (4.31), we obtain the result

$$G(\xi, x) = \begin{cases} \dfrac{1}{k}(2e^{2-2x} - 4e^{1-x})(e^\xi - e^{2\xi}), & 0 \le \xi \le x \\ \dfrac{1}{k}[(2e^{2-x} - 2e^2 - 1)\, e^{\xi - x} & \\ + (4e - 4e^{1-x} + e^{-x})\, e^{2\xi - x}], & x \le \xi \le 1 \end{cases} \tag{4.32}$$

where $k = 1 - 4e + 2e^2$. It comes as no surprise that $G(\xi, x)$ is not symmetric since the differential operator is not self-adjoint.

EXERCISES

4.4. Use the above method of piecewise integration to re-derive the Green's function (4.11) of Example 1.

4.5. Find the Green's function, and hence the solution for each of the following systems, where the interval is $0 \le x \le 1$ in each case.

[18] That the Green's function is not *always* continuous at $\xi = x$ will be seen in Exercise 4.9.

[19] Note carefully that to calculate G at $x - 0$ we use (4.24a) with ξ set equal to x, and to calculate it at $x + 0$ we use (4.24b) with ξ set equal to x.

(a) $u'' - k^2u = \phi; \qquad u(0) - u'(0) = a,\ u(1) = b$

(b) $u'' = \phi; \qquad u(0) = 0,\ u'(0) = 0$

(c) $u'' + 2u' + u = 0; \qquad u(0) = 0,\ u'(0) = 1$

4.6. For Exercise 4.5b, identify the operators $\mathscr{G}$ and $\mathscr{G}^*$ (recall Comment 3 of Example 1), compute G and G^*, and hence verify the symmetry condition (4.16) for this example.

4.7. Solve the problem $(xu')' - (1/x)u = 1$, $u(0) = u(1) = 0$, by the method of Green's functions. *Hint:* The boundary terms in the integration by parts are found to be

$$G(1, x)u'(1) - G_\xi(1, x)u(1) - G(0, x) \cdot 0 \cdot u'(0) + 0 \cdot G_\xi(0, x)u(0)$$

Since $u'(1)$ and $u'(0)$ are not prescribed, it appears that we should require $G(1, x) = G(0, x) = 0$. But due to the zero factor in the third term, we need only require that $G(1, x) = 0$ and $G(0, x) =$ finite.

Example 3. *Infinite Beam on Elastic Foundation.* Thus far, we have applied the method to second order differential operators defined over finite intervals. On the other hand, let us consider an infinitely long beam on an elastic foundation. According to the classical Euler beam theory,[20] the deflection $u(x)$ resulting from a load distribution $p(x)$ pounds per unit length satisfies the differential equation

$$(EIu'')'' = p(x) \tag{4.33}$$

where the "flexural rigidity" of the beam EI may be a function of x. We consider both EI and the "foundation modulus" k (i. e., the spring constant per unit x-length) to be constant, and the prescribed loading to be $w(x)$, as shown in Fig. 4.4. Now, $p(x)$ is the *net* loading, consisting of $w(x)$ downward, and the spring force $ku(x)$ per unit x-length upward. Thus $p(x) = w(x) - ku(x)$, and (4.33) becomes

$$EIu'''' + ku = w(x) \tag{4.34}$$

or

$$u'''' + \alpha^4 u = W(x); \qquad \text{where } \alpha^4 \equiv \frac{k}{EI} \quad \text{and} \quad W \equiv \frac{w}{EI} \tag{4.35}$$

As in the previous examples, we integrate by parts:

[20] See, for example, S. Timoshenko, *Strength of Materials*, part I, D. Van Nostrand, Princeton, N. J., 1955.

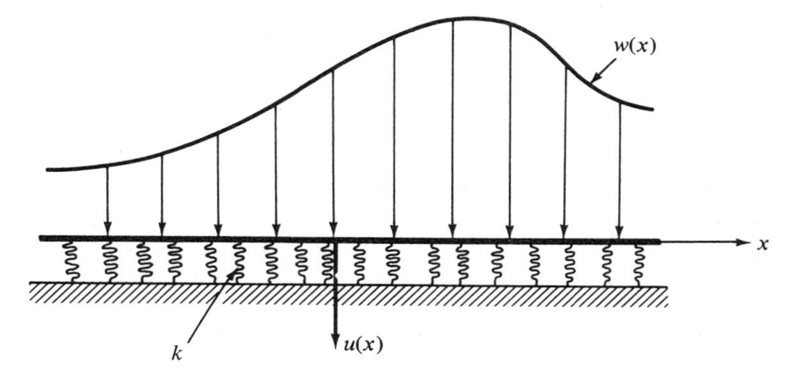

FIGURE 4.4 Infinite beam on elastic foundation.

$$\int_{-\infty}^{\infty} G\mathbf{L}u \, d\xi = \int_{-\infty}^{\infty} G(u'''' + \alpha^4 u) \, d\xi$$

$$= (Gu''' - G_\xi u'' + G_{\xi\xi} u' - G_{\xi\xi\xi} u)\Big|_{-\infty}^{\infty} + \int_{-\infty}^{\infty} u\mathbf{L}^* G \, d\xi \tag{4.36}$$

where $\mathbf{L}^* = d^4/d\xi^4 + \alpha^4 = \mathbf{L}$. Thus far we have not mentioned any boundary conditions on u. Let us merely assume that the applied loading $W(x)$ is sufficiently mild so that u, u', u'' and u''' are all *finite* at $|x| = \infty$. Then, requiring that

$$\mathbf{L}^* G = G_{\xi\xi\xi\xi} + \alpha^4 G = \delta(\xi - x) \tag{4.37}$$

together with the boundary conditions $G = G_\xi = G_{\xi\xi} = G_{\xi\xi\xi} = 0$ at $|x| = \infty$, we have the solution

$$u(x) = \int_{-\infty}^{\infty} G(\xi, x) \, W(\xi) \, d\xi \tag{4.38}$$

provided, of course, that we can compute the Green's function.

Rather than pursue a piecewise solution of (4.37), it is more convenient to leave the infinite interval intact and to Fourier transform[21] on the variable ξ. Specifically, we multiply each of the three terms in (4.37) by $\exp(-i\omega\xi) \, d\xi$ and integrate from $-\infty$ to $+\infty$. Integrating the $G_{\xi\xi\xi\xi}$ term by parts four times and applying the boundary conditions, it follows that

[21] We leave the piecewise solution for Exercise 4.10, and will use this opportunity to introduce the application of integral transform methods. The Fourier and Laplace transforms will again be used, occasionally, in PART II. A comprehensive treatment of this subject may be found in R. V. Churchill, *Operational Mathematics*, 2nd ed., McGraw-Hill Book Company, New York, 1958.

$$\omega^4 \hat{G} + \alpha^4 \hat{G} = e^{-i\omega x} \tag{4.39}$$

where the Fourier transform of G is defined by

$$\hat{G}(\omega, x) = \int_{-\infty}^{\infty} G(\xi, x)\, e^{-i\omega\xi}\, d\xi \tag{4.40}$$

Whereas (4.37) was a *differential* equation on G, the transformed equation (4.39) is an *algebraic* equation in $\hat{G}$. Solving it, we obtain

$$\hat{G} = \frac{e^{-i\omega x}}{\omega^4 + \alpha^4} \tag{4.41}$$

The well-known Fourier inversion formula then yields

$$\begin{aligned} G(\xi, x) &= \frac{1}{2\pi}\int_{-\infty}^{\infty} \frac{e^{-i\omega x}}{\omega^4 + \alpha^4} \cdot e^{i\omega\xi}\, d\omega = \frac{1}{\pi}\int_0^{\infty} \frac{\cos \omega(\xi - x)}{\omega^4 + \alpha^4}\, d\omega \\ &= \frac{e^{-\alpha|\xi - x|/\sqrt{2}}}{2\alpha^3} \sin\left(\frac{\alpha|\xi - x|}{\sqrt{2}} + \frac{\pi}{4}\right) \end{aligned} \tag{4.42}$$

the last step following upon application of the residue theorem (for those who are acquainted with complex variable theory).

COMMENT 1. Comparing (4.37) with (4.35), it is clear that $G(\xi, x)$ represents the deflection, as a function of the running variable ξ, due to a point load of unit strength acting at $\xi = x$, as sketched in Fig. 4.5.

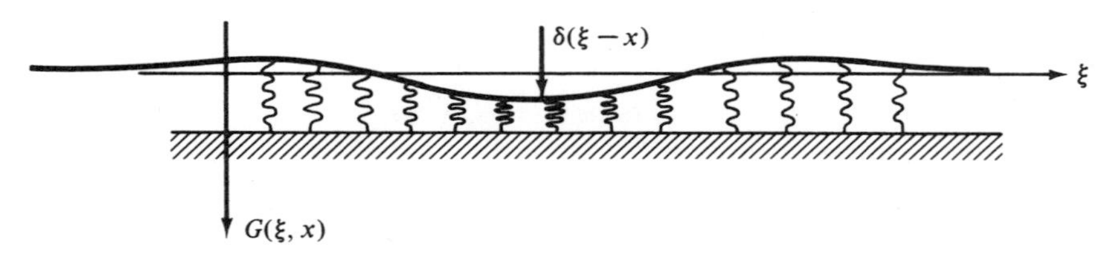

FIGURE 4.5 Green's function for Example 3.

COMMENT 2. If we *had* used piecewise integration we would have used the matching conditions G, G_ξ and $G_{\xi\xi}$ continuous at $\xi = x$, and

$$G_{\xi\xi\xi}\Big|_{x-0}^{x+0} = 1 \tag{4.43}$$

following the same reasoning as in Example 2. The reader can easily verify that these conditions are, in fact, satisfied by the Green's function (4.42). Physically (from Euler beam theory), (4.43) is a statement of the fact the "shear" jumps by unity as we pass from $\xi = x - 0$ to $x + 0$ due to the concentrated unit load at x.

COMMENT 3. Note that the singular nature of G varies from case to case. In the present example, G, G_ξ and $G_{\xi\xi}$ are all continuous functions of ξ, and it is not until we get to the third derivative $G_{\xi\xi\xi}$ that the singularity at x becomes exposed; recall (4.43). By contrast, the Green's function of Example 1 was more singular, in the sense that its *first* derivative broke down at $\xi = x$. In fact, the Green's function of Exercise 4.9 is not even continuous at $\xi = x$.

EXERCISES

4.8. Show that the solution of the initial value problem

$$\ddot{x} + \omega^2 x = f(t) \quad \text{for} \quad t \geq 0 \ [x(0) \text{ and } \dot{x}(0) \text{ prescribed}]$$

by the method of Green's functions, where the dots denote differentiation with respect to time, is

$$x(t) = \frac{\dot{x}(0)}{\omega} \sin \omega t + x(0) \cos \omega t + \int_0^\infty G(\tau, t) f(\tau)\, d\tau$$

where $G(\tau, t) = [\sin \omega(t - \tau)]/\omega$ for $\tau < t$, and 0 for $\tau > t$. Why is $G(\tau, t)$ *not* symmetric in its two arguments τ and t?

4.9. Solve the simple initial value problem

$$u' = \phi \quad \text{for} \quad x \geq 0 \ [u(0) \text{ prescribed}]$$

by the method of Green's functions. In contrast with our previous examples and exercises, we emphasize that the Green's function $G(\xi, x) = H(\xi - x) - 1$ is *not* continuous at $\xi = x$.

4.10. Obtain the Green's function (4.42), for Example 3, using the "piecewise" approach developed in Example 2.

4.11. Find the Green's function, and hence the solution, of the system

$$u'''' = \phi; \qquad u(0) = 0,\ u'(0) = 2u'(1),\ u(1) = a,\ u''(0) = 0$$

Example 4. *A Bessel Equation.* Solution of the system

$$(xu')' + k^2 xu = \phi(x); \qquad u(0) \text{ finite},\ u(\mathrm{b}) = 0 \tag{4.44}$$

where $\phi(x)$ is prescribed is no more difficult than the preceding examples, despite the nonconstant coefficients in the differential operator, and the finiteness boundary condition.

Integrating by parts, we find that

$$\int_0^b G\mathbf{L}u\,d\xi = [G(\xi, x)\,\xi u'(\xi) - G_\xi(\xi, x)\,\xi u(\xi)]\Big|_0^b + \int_0^b u\mathbf{L}^* G d\xi$$
$$= G(\mathrm{b}, x)\,\mathrm{b}u'(\mathrm{b}) - G_\xi(\mathrm{b}, x)\,\mathrm{b}u(\mathrm{b}) - G(0, x)\cdot 0\cdot u'(0)$$
$$+ G_\xi(0, x)\cdot 0\cdot u(0) + \int_0^b u\mathbf{L}^* G d\xi \tag{4.45}$$

Of the four boundary terms on the right-hand side, the second and fourth drop out, since $u(\mathrm{b}) = 0$ and $u(0)$ is finite. The first and third, however, are unwelcome since $u'(\mathrm{b})$ and $u'(0)$ are not prescribed. We therefore eliminate them by requiring that $G(0, x)$ be finite and that $G(\mathrm{b}, x) = 0$. Setting $\mathbf{L}^* G = \delta(\xi - x)$, as usual, and noting that $\mathbf{L}u = \phi$, the solution of (4.44) is therefore given by

$$u(x) = \int_0^b G(\xi, x)\,\phi(\xi)\,d\xi \tag{4.46}$$

where the Green's function satisfies the boundary value problem

$$\mathbf{L}^* G = \mathbf{L}G = (\xi G_\xi)_\xi + k^2\xi G = \delta(\xi - x) \tag{4.47}$$
$$G(0, x) \text{ finite, } G(\mathrm{b}, x) = 0 \tag{4.48}$$

To proceed with the determination of G, we split the interval into two parts, $0 \leq \xi < x$ and $x < \xi \leq \mathrm{b}$, in each of which G satisfies the homogeneous equation

$$(\xi G_\xi)_\xi + k^2\xi G = 0 \tag{4.49}$$

Now except for the presence of the k^2 factor, this would be a Bessel equation of order zero. But the k^2 can be "absorbed" by a simple stretching or contracting of the independent variable; setting $k\xi = \rho$, (4.49) reduces to the Bessel equation

$$(\rho G_\rho)_\rho + \rho G = 0 \tag{4.50}$$

the general solution of which is a linear combination of $J_0(\rho)$, the *Bessel function of first kind and order zero*, and $Y_0(\rho)$, the *Bessel function of second kind and order zero*. These admit expansions of the following form[22]

$$J_0(\rho) = \sum_{n=0}^{\infty} \frac{(-1)^n \rho^{2n}}{2^{2n}(n!)^2} \tag{4.51a}$$
$$\sim 1 \text{ as } \rho \to 0 \tag{4.51b}$$

[22] The definitive text on Bessel functions is G. N. Watson's *A Treatise on the Theory of Bessel Functions*, 2nd ed., Cambridge University Press, London, 1952.

$$Y_0(\rho) = \frac{2}{\pi}\left[\gamma + \ln\frac{\rho}{2}\right] J_0(\rho) + \frac{2}{\pi}\frac{(\rho/2)^2}{(1!)^2}$$
$$- \frac{2}{\pi}\frac{(\rho/2)^4}{(2!)^2}\left(1 + \frac{1}{2}\right) + \frac{2}{\pi}\frac{(\rho/2)^6}{(3!)^2}\left(1 + \frac{1}{2} + \frac{1}{3}\right) + \cdots \tag{4.52a}$$

$$\sim \left(\frac{2}{\pi}\right) \ln \rho \text{ as } \rho \to 0 \tag{4.52b}$$

where γ is Euler's constant 0.5772157.... Thus,

$$G(\xi, x) = \begin{cases} AJ_0(k\xi) + BY_0(k\xi), & 0 \leq \xi < x \quad (4.53a) \\ CJ_0(k\xi) + DY_0(k\xi), & x < \xi \leq \text{b} \quad (4.53b) \end{cases}$$

Observing from (4.52b) that $Y_0(k\xi)$ is infinite at $\xi = 0$, the finiteness condition on $G(0, x)$ implies that $B = 0$. The second of conditions (4.48) implies that

$$CJ_0(k\text{b}) + DY_0(k\text{b}) = 0 \tag{4.54}$$

Thus far we have only one equation, (4.54), in the three remaining unknowns A, C, D. In order to obtain two more equations we must suitably "match" the two parts of the solution, (4.53a) and (4.53b). Continuity of $G(\xi, x)$ at $\xi = x$ implies that

$$AJ_0(kx) = CJ_0(kx) + DY_0(kx) \tag{4.55}$$

Finally, we integrate (4.47) from $x - 0$ to $x + 0$:

$$\int_{x-0}^{x+0} (\xi G_\xi)_\xi \, d\xi + k^2 \int_{x-0}^{x+0} \xi G \, d\xi = 1 \tag{4.56}$$

The second integral vanishes because the integrand is bounded and the interval is infinitesimal. Integrating the first term, we obtain the jump condition

$$(\xi G_\xi)\Big|_{x-0}^{x+0} = 1 \quad \text{or} \quad G_\xi\Big|_{x-0}^{x+0} = \frac{1}{x} \tag{4.57}$$

Evaluating G_ξ at $x + 0$ from (4.53b) with ξ set equal to x, and at $x - 0$ from (4.53a) with ξ set equal to x, we obtain

$$kAJ_0'(kx) - kCJ_0'(kx) - kDY_0'(kx) = -\frac{1}{x} \tag{4.58}$$

where the primes denote differentiation with respect to the argument.

Solving (4.54), (4.55), and (4.58) for A, C, and D, we have

$$A = -\frac{1}{kx}\frac{J_0(kx)\,Y_0(kb) - J_0(kb)\,Y_0(kx)}{J_0(kb)\,[J_0(kx)\,Y_0'(kx) - J_0'(kx)\,Y_0(kx)]} \tag{4.59a}$$

$$C = -\frac{1}{kx}\frac{J_0(kx)\,Y_0(kb)}{J_0(kb)\,[J_0(kx)\,Y_0'(kx) - J_0'(kx)\,Y_0(kx)]} \tag{4.59b}$$

$$D = -\frac{1}{kx}\frac{J_0(kx)}{J_0'(kx)\,Y_0(kx) - J_0(kx)\,Y_0'(kx)} \tag{4.59c}$$

Actually, these expressions can be simplified. Since $J_0(\rho)$ and $Y_0(\rho)$ satisfy the Bessel equation (4.54), we have

$$\rho J_0'' + J_0' + \rho J_0 = 0 \tag{4.60a}$$

$$\rho Y_0'' + Y_0' + \rho Y_0 = 0 \tag{4.60b}$$

Multiplying the first by Y_0, the second by J_0, and subtracting, we obtain

$$\rho(J_0'' Y_0 - Y_0'' J_0) + (Y_0 J_0' - J_0 Y_0') = 0 \tag{4.61}$$

or

$$\frac{d}{d\rho}[\rho(Y_0 J_0' - J_0 Y_0')] = 0 \tag{4.62}$$

Thus,

$$\rho(Y_0 J_0' - J_0 Y_0') = \text{constant, for all } \rho \tag{4.63}$$

The value of the constant is most easily computed by evaluating the left-hand side, using (4.51) and (4.52), as $\rho \to 0$. We find (Exercise 4.12) that the constant is $-2/\pi$. Setting $\rho = kx$, we therefore have

$$kx[Y_0(kx)\,J_0'(kx) - J_0(kx)\,Y_0'(kx)] = -\frac{2}{\pi} \tag{4.64}$$

Using this to simplify A, C, and D, we finally obtain

$$G(\xi, x) = \begin{cases} [J_0(kb)\,Y_0(kx) - J_0(kx)\,Y_0(kb)]\,\dfrac{\pi J_0(k\xi)}{2J_0(kb)}, & 0 \leq \xi \leq x \\[2ex] [J_0(kb)\,Y_0(k\xi) - J_0(k\xi)\,Y_0(kb)]\,\dfrac{\pi J_0(kx)}{2J_0(kb)}, & x \leq \xi \leq b \end{cases} \tag{4.65}$$

provided that kb does not coincide with a zero of the Bessel function J_0. If $J_0(kb) = 0$, then the Green's function *does not exist*. This situation will be dealt with in the next example, and then further clarified toward the end of Section 5.

EXERCISES

4.12. Verify that the constant in (4.63) is $-2/\pi$.

Example 5. *The Generalized Green's Function.* As our final example, we consider the system

$$u'' + u = \phi(x); \qquad u(0) = u(\pi) = 0 \tag{4.66}$$

where $\phi(x)$ is prescribed. Although not yet obvious, this example is somewhat singular, and will require special treatment. Proceeding as before, we integrate by parts:

$$\begin{aligned}\int_0^\pi G\mathbf{L}u\,d\xi &= \int_0^\pi G(u'' + u)\,d\xi \\ &= (Gu' - G_\xi u)\Big|_0^\pi + \int_0^\pi u(G_{\xi\xi} + G)\,d\xi \\ &= G(\pi, x)\,u'(\pi) - G(0, x)\,u'(0) + \int_0^\pi u\mathbf{L}^* G\,d\xi \end{aligned}\tag{4.67}$$

and hence require of G that

$$\mathbf{L}^* G = G_{\xi\xi} + G = \delta(\xi - x); \qquad G(0, x) = G(\pi, x) = 0 \tag{4.68}$$

As in Example 2, we split the interval into two parts, $0 \leq \xi < x$ and $x < \xi \leq \pi$, in each of which $\mathbf{L}^* G = 0$. Clearly then,

$$G(\xi, x) = \begin{cases} A\sin\xi + B\cos\xi, & 0 \leq \xi < x \\ C\sin\xi + D\cos\xi, & x < \xi \leq \pi \end{cases} \tag{4.69}$$

Application of the boundary conditions implies that $B = D = 0$. To obtain the matching conditions we integrate the differential equation (4.68) from $x - 0$ to $x + 0$:

$$\int_{x-0}^{x+0} (G_{\xi\xi} + G)\,d\xi = \int_{x-0}^{x+0} \delta(\xi - x)\,d\xi \tag{4.70}$$

Requiring G to be continuous at x, it follows that

$$A = C \tag{4.71}$$

and (4.70) reduces to the jump condition

$$G_\xi\Big|_{x-0}^{x+0} = 1 \tag{4.72}$$

or

$$C\cos x - A\cos x = 1 \tag{4.73}$$

Unfortunately, equations (4.71) and (4.73) in the two unknowns A and C are incompatible, so that our "usual" Green's function *does not exist*![23]

To get at the source of this difficulty, consider for a moment the *homogeneous* version of (4.68),

$$v_{\xi\xi} + v = 0; \qquad v(0) = v(\pi) = 0 \tag{4.74}$$

[23] Actually, this situation does not arise very often in practice. Nevertheless, it is well worth discussing, not only for the sake of completeness, but also because some important general features of linear equations will be brought out in the discussion.

This admits nontrivial solutions which are arbitrary multiples of $\sin \xi$. "Dotting" v into equation (4.68), i.e., multiplying each term by $v(\xi) = \sin \xi$ and integrating from 0 to π,

$$\int_0^\pi v(G_{\xi\xi} + G)\, d\xi = \int_0^\pi v\delta(\xi - x)\, d\xi \tag{4.75}$$

$$(vG_\xi - v_\xi G)\Big|_0^\pi + \int_0^\pi G(v_{\xi\xi} + v)\, d\xi = v(x) \tag{4.76}$$

so that, recalling (4.74) and the boundary conditions on G, we have the contradiction

$$0 + 0 = v(x) \tag{4.77}$$

One might say, then, that the difficulty is due to the fact that $v(\xi)$ is not "orthogonal" to $\delta(\xi - x)$, that is, the right-hand side of (4.75) is not zero. This provides us with a clue as to how we can patch things up. Instead of requiring that G satisfy the system (4.68), let us require that

$$\mathbf{L}^*G = G_{\xi\xi} + G = \delta(\xi - x) + F; \qquad G(0, x) = G(\pi, x) = 0 \tag{4.78}$$

where F is chosen such that $v(\xi) = \sin \xi$ *is* orthogonal to the combination $\delta(\xi - x) + F$. We will now demonstrate that a suitable general form for F is $F = \alpha v(x)\, v(\xi)$. That is, the constant α can in fact be chosen so as to obtain the desired orthogonality, since

$$\int_0^\pi v(\xi)\, [\delta(\xi - x) + \alpha v(x)\, v(\xi)]\, d\xi = v(x)\left[1 + \alpha \int_0^\pi v^2(\xi)\, d\xi\right] = 0 \tag{4.79}$$

if

$$\alpha = -\frac{1}{\int_0^\pi v^2(\xi)\, d\xi} \tag{4.80}$$

In the present example $v(\xi) = \sin \xi$ so that $\alpha = -2/\pi$.

Before proceeding with the solution of (4.78) for the modified or *generalized* Green's function G, we note that the original system (4.66) is of the same form as (4.68). Since there is no solution of (4.68), because $\delta(\xi - x)$ is not orthogonal to the homogeneous solution $v(\xi) = \sin \xi$, identical reasoning tells us that there will be no solution to (4.66) unless $\phi(x)$ is orthogonal to $v(x) = \sin x$. For our original problem to be well-posed, then, we will restrict the allowable $\phi(x)$'s to be such that

$$\int_0^\pi \phi(x) \sin x\, dx = 0 \tag{4.81}$$

Returning to our calculation of $G(\xi, x)$ from (4.78) we find easily that

$$G(\xi, x) = \frac{(\sin x)(\xi \cos \xi)}{\pi} + \begin{cases} A \sin \xi + B \cos \xi, & 0 \le \xi < x \\ C \sin \xi + D \cos \xi, & x < \xi \le \pi \end{cases} \tag{4.82}$$

where the first term on the right-hand side is the particular solution due to our additional $-(2/\pi) \sin x \sin \xi$ term in the differential equation. The boundary conditions are

$$G(0, x) = 0 = B \tag{4.83}$$

$$G(\pi, x) = 0 = -\sin x - D \tag{4.84}$$

Integrating (4.78) from $\xi = x - 0$ to $\xi = x + 0$, we find that continuity of G at $\xi = x$, together with the jump condition (4.72), constitutes suitable matching conditions. In terms of our unknown coefficients

$$A \sin x + B \cos x = C \sin x + D \cos x \tag{4.85}$$

and

$$C \cos x - D \sin x - A \cos x + B \sin x = 1 \tag{4.86}$$

respectively. Equations (4.83)–(4.86) are, in fact, compatible and admit the solution

$$\begin{aligned} A &= C - \cos x \\ B &= 0 \\ C &= \text{arbitrary function of } x \\ D &= -\sin x \end{aligned} \tag{4.87}$$

so that

$$G(\xi, x) = C(x) \sin \xi + \frac{(\sin x)(\xi \cos \xi)}{\pi} - \begin{cases} \cos x \sin \xi, & 0 \le \xi \le x \\ \sin x \cos \xi, & x \le \xi \le \pi \end{cases} \tag{4.88}$$

With G determined, the solution is obtained from (4.67), with the help of (4.66) and (4.78), as follows:

$$\int_0^\pi G(\xi, x)\, \phi(\xi)\, d\xi = u(x) - \frac{2}{\pi} \sin x \int_0^\pi u(\xi) \sin \xi \, d\xi \tag{4.89}$$

Since the integral on the right is a constant, we can rewrite this as

$$u(x) = K \sin x + \int_0^\pi G(\xi, x)\, \phi(\xi)\, d\xi \tag{4.90}$$

The constant K is arbitrary for the same reason that $C(x)$ was arbitrary in (4.88), namely, $\sin x$ is a solution of the *homogeneous* version of (4.66), just as $\sin \xi$ was a solution of the homogeneous version of (4.78); the same is true for an arbitrary multiple of $\sin x$. Notice that $C(x)$ has no effect upon the solution (4.90) since the term

$$\int_0^\pi C(x) \sin \xi \phi(\xi)\, d\xi = C(x) \int_0^\pi \sin \xi \phi(\xi)\, d\xi \tag{4.91}$$

is zero anyway, by virtue of the restriction (4.81).

It may be helpful to summarize briefly the special features of the above example. We found that the system (4.68) governing G did not admit a solution—because the right-hand side $\delta(\xi - x)$ was not orthogonal to the solution $\sin \xi$ of the homogeneous version of (4.68). This was remedied by adding a suitable function F to the right-hand side such that the combination $\delta(\xi - x) + F$ did in fact satisfy the desired orthogonality.

COMMENT 1. In the above example there was only one nontrivial solution of the homogeneous system $\mathbf{L}^* G = 0$, $G(0, x) = G(\pi, x) = 0$; namely, $v(\xi) = \sin \xi$. If there had been two such (linearly independent) solutions, say $v_1(\xi)$ and $v_2(\xi)$, then we would have required that

$$\mathbf{L}^* G = \delta(\xi - x) + \alpha_1 v_1(x)\, v_1(\xi) + \alpha_2 v_2(x)\, v_2(\xi) \tag{4.92}$$

with α_1 and α_2 chosen so that

$$\int_0^\pi v_j(\xi)\, [\delta(\xi - x) + \alpha_1 v_1(x)\, v_1(\xi) + \alpha_2 v_2(x)\, v_2(\xi)]\, d\xi = 0 \tag{4.93}$$

for $j = 1$ and 2.

COMMENT 2. In dealing with our system (4.66) or, to make it a little more general,

$$u'' + p(x)\, u = \phi(x); \qquad u(0) = u(\pi) = 0 \tag{4.94}$$

we took $\mathbf{L}$ to be $d^2/dx^2 + p(x)$. Alternatively, we could have taken $\mathbf{L} = d^2/dx^2$, that is,

$$\mathbf{L}u = u'' = \phi(x) - p(x)\, u \tag{4.95}$$

Now, the Green's function for the operator $\mathbf{L} = d^2/dx^2$ and the boundary conditions $u(0) = u(\pi) = 0$ is (as the reader can verify) simply

$$G(\xi, x) = (\xi - x)\, H(\xi - x) + \frac{x\xi}{\pi} - \xi$$
$$= \begin{cases} \dfrac{(x - \pi)\,\xi}{\pi}, & 0 \le \xi \le x \\ \dfrac{(\xi - \pi)\,x}{\pi}, & x \le \xi \le \pi \end{cases} \tag{4.96}$$

and the solution of (4.94) is

$$u(x) = \int_0^\pi G(\xi, x)\, \phi(\xi)\, d\xi - \int_0^\pi G(\xi, x)\, \rho(\xi)\, u(\xi)\, d\xi \tag{4.97}$$

Actually, this is not the "solution" since u appears inside the second integral. It is therefore an *integral equation* on u; to be precise, a Fredholm equation of the second kind, with the kernel $G(\xi, x)\, \rho(\xi)$. Whereas $G(\xi, x)$ is symmetric (in ξ and x), the kernel $G\rho$ won't be unless ρ is a constant. Since symmetry of the kernel is necessary for the application of the powerful Hilbert-Schmidt theory,[24] it is helpful to convert (4.97) to an equation with a symmetric kernel. If $\rho(x) \ge 0$ over the interval, this can be accomplished by multiplying each term in (4.97) by $\sqrt{\rho(x)}$. This produces the integral equation

$$v(x) = F(x) + \int_0^\pi K(\xi, x)\, v(\xi)\, d\xi \tag{4.98}$$

in the new unknown $v(x) = \sqrt{\rho(x)}\, u(x)$, where

$$F(x) = \sqrt{\rho(x)} \int_0^\pi G(\xi, x)\, \phi(\xi)\, d\xi \tag{4.99}$$

$$K(\xi, x) = -\sqrt{\rho(x)\, \rho(\xi)}\, G(\xi, x) = K(x, \xi) \tag{4.100}$$

In the case of especially complicated boundary value problems, where an exact solution is apparently out of the question, it is often best to recast the problem in the form of an integral equation, since integral equations are generally better vehicles for methods of approximation. Roughly speaking, the idea is that integration is a "smoothing" operation, whereas the effect of a differential operator is just the opposite.

This idea applies also in the case of *nonlinear* systems. For example, the equation $\mathbf{L}u = \phi(x, u)$, where ϕ is a nonlinear function of u, is generally quite intractable. The best line of approach might be to regard ϕ as known, and solve the equation by means of Green's functions. Because of the u dependence in ϕ, the "solution" will actually be a nonlinear integral equation, which may very

[24] See, for example, R. Courant and D. Hilbert, *Methods of Mathematical Physics*, Vol. 1, Interscience Publishers, Inc., New York, 1953.

well be susceptible to various methods of approximation, such as the method of successive approximations or Newton's method.

EXERCISES

4.13. As another example of the generalized Green's function, show that the solution of the system

$$\mathbf{L}u = u'' = \phi(x); \qquad u'(0) = u'(1) = 0$$

is given by

$$u(x) = K + \int_0^1 (\xi - x)\, H(\xi - x)\, \phi(\xi)\, d\xi = K + \int_x^1 (\xi - x)\, \phi(\xi)\, d\xi$$

where K is an arbitrary constant and $\phi(x)$ is restricted so that

$$\int_0^1 \phi(x)\, dx = 0$$

4.14. In Exercise 4.13 we found that an ordinary Green's function, satisfying $G_{\xi\xi} = \delta(\xi - x)$ and $G_\xi(0, x) = G_\xi(1, x) = 0$, does not exist. We remedied this, as in Example 5, by replacing $\delta(\xi - x)$ by $\delta(\xi - x) + F$. Alternatively, let us patch things up by leaving $\delta(\xi - x)$ intact, and changing instead the boundary conditions on G. First, obtain

$$u(x) = u(1)\, G_\xi(1, x) - u(0)\, G_\xi(0, x) + \int_0^1 G(\xi, x)\, \phi(\xi)\, d\xi \tag{A}$$

as usual, where $G_{\xi\xi} = \delta(\xi - x)$, but boundary conditions on G have not yet been specified. Show that if we impose the condition that

$$G_\xi(1, x), \qquad G_\xi(0, x), \qquad G_\xi(\xi, 0), \qquad G_\xi(\xi, 1) \tag{B}$$

all are constant, then (A) *does* satisfy all the conditions required of $u(x)$. Then show that a "modified-generalized Green's function" (for lack of a better name) satisfying $G_{\xi\xi} = \delta(\xi - x)$ plus conditions (B) *does* exist, and is given by

$$G(\xi, x) = (\xi - x)\, H(\xi - x) + A\xi + B(x)$$

where the constant A and the function $B(x)$ are arbitrary. Finally, show that this yields the solution stated in Exercise 4.13.

5. THE EIGENFUNCTION METHOD

Basically, there are two different lines of approach to the problem

$$\mathbf{L}u = \phi \tag{5.1}$$

One approach is to seek the *inverse operator*, say $\mathbf{L}^{-1}$, such that

$$u = \mathbf{L}^{-1}\phi \tag{5.2}$$

This is, in fact, the Green's function method; $\mathbf{L}^{-1}$ turns out to be an integral operator, the kernel of which is the Green's function.

The other line of approach is the eigenfunction method. Since this method is actually an *alternative* to the inverse operator, or Green's function method, and requires a good deal of theoretical development, we will limit out discussion here to only a brief introduction, partly for the sake of comparison and partly because it will shed further light on the *generalized* Green's function already discussed in Section 4.

The Eigenvalue Problem. To introduce the basic ideas of the eigenfunction method, let us consider the simple boundary value problem

$$u'' + \lambda u = 0; \qquad u(0) = u(\ell) = 0 \tag{5.3}$$

where λ is a parameter. Clearly $u(x) = 0$ is a solution of this system for *all* values of λ. Nevertheless, let us try to find *non*trivial solutions. The general solution of the differential equation is

$$u(x) = A \sin \sqrt{\lambda}\, x + B \cos \sqrt{\lambda}\, x \tag{5.4}$$

and the boundary condition $u(0) = 0$ implies that $B = 0$. Applying the second boundary condition

$$u(\ell) = 0 = A \sin \sqrt{\lambda}\, \ell \tag{5.5}$$

If we satisfy this by setting $A = 0$, we obtain the trivial solution $u(x) = 0$. On the other hand, (5.5) will also be satisfied if the parameter λ is such that $\sqrt{\lambda}\, \ell$ coincides with a zero of the sine function; namely, if

$$\sqrt{\lambda}\, \ell = n\pi, \qquad \text{so that } \lambda = \frac{n^2\pi^2}{\ell^2} \equiv \lambda_n \tag{5.6}$$

for $n = 1, 2, 3, \ldots$. In that case we have the nontrivial solutions

$$u(x) = A \sin\left(\frac{n\pi x}{\ell}\right) \equiv A\phi_n(x) \text{ say} \tag{5.7}$$

where A is an arbitrary constant factor. Note that we exclude the case $n = 0$; $\lambda = 0$ is *not* an eigenvalue in this example (although in other cases it may be, as in Exercise 5.7) because $\phi_0(x) = 0$, thus yielding only the trivial solution $u(x) = 0$.

We call the special values $\lambda_n = n^2\pi^2/\ell^2$ the characteristic values or **eigenvalues,** and the corresponding functions $\phi_n(x) = \sin(n\pi x/\ell)$ the characteristic functions or **eigenfunctions** of the *eigenvalue problem* (5.3).

COMMENT 1. Notice that the nonuniqueness of the solution, when λ coincides with an eigenvalue, does *not* violate any of the standard existence-uniqueness theorems, since these theorems apply to *initial* value problems. Surely, $u = 0$ *would* be the unique solution if we replaced the right-hand boundary condition $u(\ell) = 0$ in (5.3) by the condition $u'(0) = 0$.

COMMENT 2. When eigenvalue problems such as (5.3) arise in connection with physical situations, the eigenvalues and eigenfunctions are generally of considerable physical interest. For example, suppose we apply axial forces P to the ends of a straight column (Fig. 5.1). It turns out that the resulting lateral deflection $u(x)$ is governed by the boundary value problem (5.3), where $\lambda = P/EI$. The product EI is a physical constant of the column, called the "flexural rigidity," which is a measure of the column's resistance to flexure.

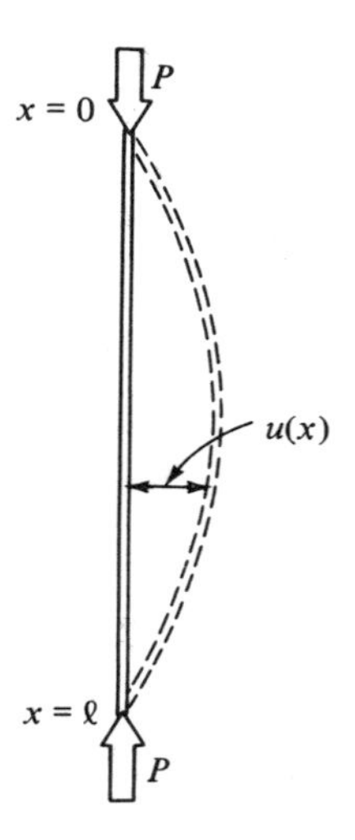

FIGURE 5.1 Buckling of a column.

According to our above analysis, we will have $u(x) = 0$ *unless* the end force P is such that $\lambda = P/EI = n^2\pi^2/\ell^2$, in which case the deflection is given by (5.7). Physically, $P_n = n^2\pi^2 EI/\ell^2$ are the "buckling loads" and $\phi_n(x) = \sin(n\pi x/l)$ are the corresponding "buckling modes." For design purposes, the lowest buckling load P_1 is of most interest and is called the "critical" buckling load, P_{cr}. (The fact that $P_{cr} \to \infty$ as $\ell \to 0$, and $P_{cr} \to 0$ as $\ell \to \infty$ is in agreement with our experience and intuition.)

COMMENT 3. It is a striking fact that the eigenfunctions $\phi_n = \sin(n\pi x/\ell)$ generated by the eigenvalue problem (5.3) are precisely the quantities needed for Fourier series representation of functions defined over $0 \leq x \leq \ell$. In a sense, this is where the terms of a Fourier sine series "come from."

The fact is that (5.3) is a special case of the more general **Sturm-Liouville** (*not* Louisville) eigenvalue problem consisting of the differential equation

$$\mathbf{L}u + \lambda\, ru = (pu')' + qu + \lambda\, ru = 0 \tag{5.8}$$

over $\mathrm{a} \leq x \leq \mathrm{b}$, plus homogeneous boundary conditions of the type

$$\mathbf{B}_1(u) = a_1 u(\mathrm{a}) + a_2 u'(\mathrm{a}) = 0 \tag{5.9a}$$

$$\mathbf{B}_2(u) = b_1 u(\mathrm{b}) + b_2 u'(\mathrm{b}) = 0 \tag{5.9b}$$

If we assume further that $p(x)$, $q(x)$, and $r(x)$ have continuous derivatives[25] and $p(x) > 0$, $r(x) > 0$ over the interval, and if we denote the eigenfunctions of the Sturm-Liouville system (5.8)–(5.9) by $\phi_n(x)$, then it can be shown that the functions $\sqrt{r}\,\phi_n$ $(n = 1, 2, \dots)$ are mutually **orthogonal,** that is,

$$(\sqrt{r}\,\phi_m, \sqrt{r}\,\phi_n) = \int_{\mathrm{a}}^{\mathrm{b}} \sqrt{r}\,\phi_m \sqrt{r}\,\phi_n\, dx = 0, \qquad \text{for } m \neq n \tag{5.10}$$

Furthermore, the orthogonal set $\{\sqrt{r}\,\phi_n\}$ is **complete,** in the sense of **mean square convergence.** That is, suppose we construct the **Fourier series** of a given function $f(x)$, namely

$$\sum_{n=1}^{\infty} c_n \sqrt{r}\,\phi_n \tag{5.11}$$

where the so-called *Fourier coefficients* are defined by

$$c_n = \frac{(f, \sqrt{r}\,\phi_n)}{(\sqrt{r}\,\phi_n, \sqrt{r}\,\phi_n)} \tag{5.12}$$

Then, if f is reasonably well-behaved (e. g., piecewise continuous[26]), the limit of the mean square error vanishes,

[25] It is shocking to learn that the derivative of a function may exist, and yet *not* be continuous. [For a function $f(x)$ to be continuous at a given point $x = a$, we need both $f(a)$ and $\lim_{x \to a} f(x)$ to exist and to be equal.] For example, consider the function $g(x)$ defined by $x^2 \sin(1/x)$ for $x \neq 0$, and zero for $x = 0$. (Note that a definite value must be defined for $x = 0$ since $\sin \infty$ is meaningless.) From the definition of derivative, we compute $g'(0) = \lim_{h \to 0} [(h^2 \sin(1/h) - 0)/h] = 0$ since $\sin(1/h)$ is bounded as $h \to 0$. On the other hand, for $x \neq 0$ we have $g'(x) = 2x \sin(1/x) - \cos(1/x)$ and, because of the $\cos(1/x)$ term, $\lim_{x \to 0} g'(x)$ does not exist. Thus, even though $g'(x)$ exists at $x = 0$, it is not continuous there.

[26] We say that a function is *piecewise continuous* over a certain interval if the interval can be subdivided into a finite number of subintervals, inside each of which the function is continuous with finite limits at both endpoints.

$$\lim_{N\to\infty} \int_a^b [f(x) - \sum_{n=1}^{N} c_n \sqrt{r(x)}\, \phi_n(x)]^2\, dx = 0 \tag{5.13}$$

It is to be noted that mean square convergence (5.13)does *not* imply that

$$\lim_{N\to\infty} \sum_{n=1}^{N} c_n \sqrt{r(x)}\, \phi_n(x) = f(x) \tag{5.14}$$

for each fixed x in the interval (recall Exercise 3.5), i.e., **pointwise convergence.** Nevertheless, positive statements concerning pointwise convergence can be made if the class of f's is suitably restricted.

Application of Eigenfunction Method. Consider the problem

$$\mathbf{L}u = \phi \tag{5.15}$$

subject to boundary conditions of the type (5.9a, b), where $\mathbf{L}$ is a formally self-adjoint second order ordinary differential operator. To solve this by the eigenfunction method we consider the associated Sturm-Liouville eigenvalue problem

$$\mathbf{L}u + \lambda u = 0 \tag{5.16}$$

with the same boundary conditions as (5.15), where we have set $r(x) = 1$ for simplicity. Our procedure is to expand the various quantities in (5.15) in terms of the eigenfunctions ϕ_n of (5.16):

$$u(x) = \sum_{n=1}^{\infty} a_n \phi_n(x) \tag{5.17}$$

$$\phi(x) = \sum_{n=1}^{\infty} c_n \phi_n(x) \tag{5.18}$$

where the a_n's are unknown, and the c_n's are computed, according to (5.12), by the formula

$$c_n = \frac{(\phi, \phi_n)}{(\phi_n, \phi_n)} \tag{5.19}$$

Inserting (5.17) and (5.18) into (5.15),

$$\mathbf{L} \sum_{n=1}^{\infty} a_n \phi_n = \sum_{n=1}^{\infty} c_n \phi_n \tag{5.20}$$

Proceeding formally,[27] the left-hand side becomes

$$\mathbf{L} \sum_{n=1}^{\infty} a_n \phi_n = \sum_{n=1}^{\infty} a_n \mathbf{L}\phi_n = -\sum_{n=1}^{\infty} a_n \lambda_n \phi_n \tag{5.21}$$

[27] $(d/dx) \sum^{\infty} f_n(x) = \sum^{\infty} f_n'(x)$ only if the f_n's are suitably restricted. In (5.21), on the other hand, we have interchanged the order of application of $\mathbf{L}$ and $\sum^{\infty}$ "formally"; i. e., without worrying about the precise conditions under which the interchange is rigorous.

where the last step follows from (5.16). Now, (5.20) and (5.21) imply that

$$\sum_{n=1}^{\infty} (a_n \lambda_n + c_n)\, \phi_n = 0 \tag{5.22}$$

and, since the ϕ_n's are orthogonal, it follows (Exercise 5.3) that

$$a_n \lambda_n + c_n = 0 \tag{5.23a}$$

$$a_n = -\frac{c_n}{\lambda_n} \tag{5.23b}$$

for each n, so that we obtain

$$u(x) = -\sum_{n=1}^{\infty} \left(\frac{c_n}{\lambda_n}\right) \phi_n(x) \tag{5.24}$$

Thus, the eigenfunction solution of (5.15) consists of first determining the eigenvalues λ_n and eigenfunctions $\phi_n(x)$ of the associated Sturm-Liouville problem (5.16), then computing the Fourier coefficients c_n of $\phi(x)$ with respect to these eigenfunctions using (5.19), and finally inserting these various quantities into (5.24).

The connection between the eigenfunction expansion (5.24) and Green's functions can be put in evidence if we insert

$$c_n = \frac{1}{(\phi_n, \phi_n)} \int_a^b \phi_n(\xi)\, \phi(\xi)\, d\xi \tag{5.25}$$

into (5.24). Formally interchanging the order of integration and summation, we arrive at the form

$$u(x) = \int_a^b \left\{ -\sum_{n=1}^{\infty} \frac{\phi_n(\xi)\, \phi_n(x)}{\lambda_n(\phi_n, \phi_n)} \right\} \phi(\xi)\, d\xi \tag{5.26}$$

and, based upon our experience with Green's functions, we identify the kernel as the Green's function

$$-\sum_{n=1}^{\infty} \frac{\phi_n(\xi)\, \phi_n(x)}{\lambda_n(\phi_n, \phi_n)} = G(\xi, x) \tag{5.27}$$

To illustrate the correspondence (5.27), let us return to the loaded string example of Section 4. The boundary value problem is

$$u''(x) = \phi(x); \qquad u(0) = u(1) = 0 \tag{5.28}$$

and the associated Sturm-Liouville problem is

$$u'' + \lambda u = 0; \qquad u(0) = u(1) = 0 \tag{5.29}$$

Since this is identical to the Sturm-Liouville system (5.3), with $\ell = 1$, we have

$$\lambda_n = n^2 \pi^2, \qquad \phi_n(x) = \sin n\pi x \tag{5.30}$$

and

$$(\phi_n, \phi_n) = \int_0^1 \sin^2 n\pi x \, dx = \frac{1}{2} \tag{5.31}$$

so that, according to (5.27),

$$G(\xi, x) = -\sum_{n=1}^{\infty} \frac{2 \sin n\pi\xi \sin n\pi x}{n^2\pi^2} \tag{5.32}$$

On the other hand, we found in Section 4 that

$$G(\xi, x) = \begin{cases} (x-1)\xi, & \xi \leq x \\ (\xi - 1)x, & \xi \geq x \end{cases} \tag{5.33}$$

These two forms of G are, of course, in agreement since (5.32) is simply the Fourier sine series expansion of (5.33) over $0 \leq \xi \leq 1$!

Finally, we have two more comments:

COMMENT 4. We see, from the expression (5.27), that the Green's function fails to exist in the event that one of the eigenvalues of the associated Sturm-Liouville system, say λ_j, is *zero*! Furthermore, we see from (5.24) that the *solution* $u(x)$ fails to exist if $\lambda_j = 0$ and $c_j \neq 0$. If $c_j = 0$ also, however, then $c_j/\lambda_j = 0/0$ is indeterminate. To resolve this case we go back to (5.23a); if $\lambda_j = c_j = 0$ then (5.23a) is satisfied for *arbitrary* a_j, and (5.24) may be rewritten as

$$u(x) = K\phi_j(x) - \sum_{\substack{n=1 \\ (n \neq j)}}^{\infty} \left(\frac{c_n}{\lambda_n}\right)\phi_n(x) \tag{5.34}$$

where K is an arbitrary constant. Summarizing, *we have seen that if zero is not an eigenvalue of (5.16), then both the Green's function G and the solution u exist, and are given by (5.27) and (5.24) respectively. If, on the other hand, some* $\lambda_j = 0$, *then the (ordinary) Green's function fails to exist. As for the solution u, the situation depends on the jth Fourier coefficient of* ϕ, *namely* $c_j = (\phi, \phi_j)$. *There are two possibilities: (i) If* $c_j \neq 0$, *then a solution for u fails to exist. (ii) If* $c_j = 0$, *then an infinity of solutions exist, given by (5.34).*

With these ideas in mind, it would probably be a good idea to read Example 5 of Section 4 through once more.

COMMENT 5. Observe that homogeneity of the boundary conditions is basic to the eigenfunction method. However, this is not really any cause for excitement because we can generally "*homogenize*" the boundary conditions, The system

$$u'' + u' + xu = \phi; \qquad u(0) = 2, u(1) = 0 \tag{5.35}$$

(for example) is converted to

$$v'' + v' + xv = \psi; \qquad v(0) = v(1) = 0 \tag{5.36}$$

where $\psi = \phi + 2 + 2x(x - 1)$, by the simple change of variables $v = u + 2(x - 1)$.

EXERCISES

5.1. Show that the operator $\mathscr{L}$ of the Sturm-Liouville system is self-adjoint; $\mathscr{L}$ consists of $\mathbf{L}$ from (5.8), and $\mathbf{B}_1$, $\mathbf{B}_2$ from (5.9).

5.2. Verify the statement of orthogonality, (5.10). *Hint:* Multiply the equations $\mathbf{L}\phi_m + \lambda_m r\phi_m = 0$ and $\mathbf{L}\phi_n + \lambda_n r\phi_n = 0$ by ϕ_n and ϕ_m respectively, subtract them, and then integrate them over the interval. Thus deduce that $(\sqrt{r}\,\phi_m, \sqrt{r}\phi_n) = 0$ provided that $\lambda_m \neq \lambda_n$. [Under the assumptions that we have made on p, q, r it can be shown, further, that the Sturm-Liouville problem cannot have two linearly independent eigenfunctions corresponding to the same eigenvalue; see, for example, Churchill's *Fourier Series and Boundary Value Problems*, 2nd Edition, Section 34, Theorem 5. Thus it follows that if ϕ_m and ϕ_n are indeed independent eigenfunctions then λ_m and λ_n *must* be distinct, thus producing the statement (5.10).]

5.3. Verify that (5.22) does imply (5.23a).

5.4. Show that the eigenfunctions of the Sturm-Liouville system $u'' + \lambda u = 0$, $u(0) = 0$, $u(1) - 2u'(1) = 0$ are $\phi_n = \sin\sqrt{\lambda_n}\,x$, where the eigenvalues are solutions of the transcendental equation $\tan\sqrt{\lambda} = 2\sqrt{\lambda}$. Show that the root $\lambda = 0$ is *not* an eigenvalue. By graphical examination of the transcendental equation, deduce the asymptotic distribution of eigenvalues, $\lambda_n \sim (2n - 1)^2\pi^2/4$ as $n \to \infty$. Verify the orthogonality of the eigenfunctions, for this particular example, by direct integration of the integral in (5.10). Obtain the Fourier expansion of the function $f(x) = x$, over the interval $0 \leq x \leq 1$, in terms of the $\sin\sqrt{\lambda_n}\,x$ eigenfunctions. *Observe that it differs from the "usual" Fourier sine series in the sense that it is not harmonic; that is, the "frequencies" are not simple multiples of some fundamental frequency.*

5.5. Use the results of Exercise 5.4, together with (5.24), to obtain the eigenfunction solution of the simple boundary value problem $u'' = x$, $u(0) = 0$, $u(1) - 2u'(1) = 0$.

5.6. Find the eigenvalues and eigenfunctions of the system $(xu')' + \lambda xu = 0$, $u(0)$ bounded, $u(1) = 0$.

5.7. Show that $\lambda = 0$ is an eigenvalue of the Sturm-Liouville system $(xu')' + \lambda(x^2 + 3)u = 0$, $u'(0) = u'(1) = 0$. What is the corresponding eigenfunction?

6. SUMMARY

Before moving on to partial differential equations, let us briefly recap the basic features of the Green's function method, as we have presented it.

We have considered boundary value problems of the form

$$\mathbf{L}u = \phi(x); \qquad \mathbf{B}_j(u) = c_j \quad \text{for} \quad j = 1, 2, \ldots, n \tag{6.1}$$

over some interval $a \leq x \leq b$, not necessarily bounded, where $\mathbf{L}$ is an nth order linear ordinary differential operator and the $\mathbf{B}_j$'s are linear combinations of u and its first $n - 1$ derivatives at the endpoints of the interval. The boundary conditions need not be homogeneous, i. e., the c_j's need not all be zero.

We start by forming the inner product $(G, \mathbf{L}u)$ and integrating by parts:

$$(G, \mathbf{L}u) = \text{boundary terms} + (u, \mathbf{L}^* G) \tag{6.2}$$

Noting that $\mathbf{L}u = \phi$, and setting

$$\mathbf{L}^* G = \delta(\xi - x) \tag{6.3}$$

(6.2) becomes

$$u(x) = \text{boundary terms} + \int_a^b G(\xi, x)\, \phi(\xi)\, d\xi \tag{6.4}$$

Specification of the boundary value problem on G is completed by requiring that G satisfy homogeneous boundary conditions which remove any unwelcome boundary terms from (6.4), i.e., terms involving unspecified boundary values of u and its derivatives.

Once G is found, by solving (6.3) together with these homogeneous boundary conditions, the solution of (6.1) is given by (6.4).

In the event that the Green's function does not exist, we may remedy the situation either by changing (6.3) to

$$\mathbf{L}^* G = \delta(\xi - x) + \text{a suitably chosen function} \tag{6.5}$$

or by retaining (6.3) and adjusting the boundary conditions imposed on G (Exercise 4.14).

Under certain conditions on the operator, it may, alternatively, be possible to obtain G in the form of an eigenfunction expansion, as discussed in Section 5.

II. Application to Partial Differential Equations

To permit a reasonably systematic development, we restrict our discussion, in the first two sections of PART II, to the case of second order equations in two independent variables. Several examples which are not of this type are then presented in Section 11 for completeness, and it is shown that their solution by the method of Green's functions follows easily from the methods already developed in the preceding sections. The overall development parallels that of PART I, except for the introduction of "principal solutions" to facilitate the calculations.

1. INTRODUCTION

We consider here the general linear partial differential equation of second order, with two independent variables x and y (sometimes x and t),

$$\mathbf{L}u = Au_{xx} + 2Bu_{xy} + Cu_{yy} + Du_x + Eu_y + Fu = \phi \tag{1.1}$$

where $A, \cdots, F, \phi$ are given functions of x and y, subscripted variables denote partial differentiation, and the 2 is introduced for convenience. Equation (1.1) is to hold over some prescribed region $\mathscr{S}$ of the x, y plane, which is not necessarily bounded; see Fig. 1.1.

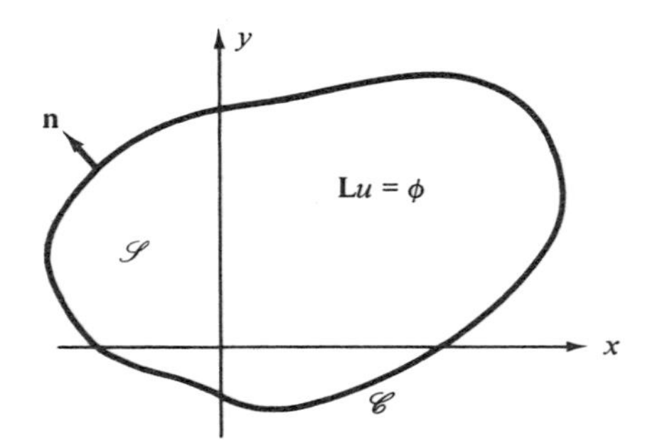

FIGURE 1.1 The boundary value problem.

In addition, linear boundary conditions of the general form

$$\mathbf{B}(u) = \alpha u + \beta u_n = f \tag{1.2}$$

are prescribed over the boundary curve $\mathscr{C}$ of the region, where u_n denotes the derivative $\partial u/\partial n$ in the direction of the outward normal, and the quantities α, β, f may vary along $\mathscr{C}$. As in PART I, the boundary conditions need *not* be homogeneous.

It is customary to classify $\mathbf{L}$ as **elliptic, parabolic,** or **hyperbolic** depending on whether the quantity $B^2 - AC$ is negative, zero, or positive, respectively. This distinction is crucial in determining the fundamental features of the solution.

To illustrate this point, consider briefly a basic problem of aerodynamics: the calculation of the air disturbance caused by an airfoil moving at a steady speed U through otherwise undisturbed air. It is best to view the flow from a coordinate system which is fixed to the airfoil, since this reduces the problem to a steady (i.e., time independent) one, as illustrated in Fig. 1.2. It can be shown that the velocity potential[1] u is governed approximately by the partial differential equation

$$(1 - M^2)u_{xx} + u_{yy} = 0 \tag{1.3}$$

where the "Mach number" M is the ratio of U to the speed of sound. Now, for the "subsonic" case, where $M < 1$, we have $B^2 - AC = M^2 - 1 < 0$, the equation (1.3) is elliptic,[2] and the flow is quite "smooth,"[3] as sketched in Fig. 1.2(a). For the "supersonic" case however, where $M > 1$, the equation (1.3) becomes hyperbolic, and the flow exhibits a wave-like structure, as seen in Fig. 1.2(b), with discontinuities in the normal velocity component (and pressure) across the "shock waves" *abc* and *def*!

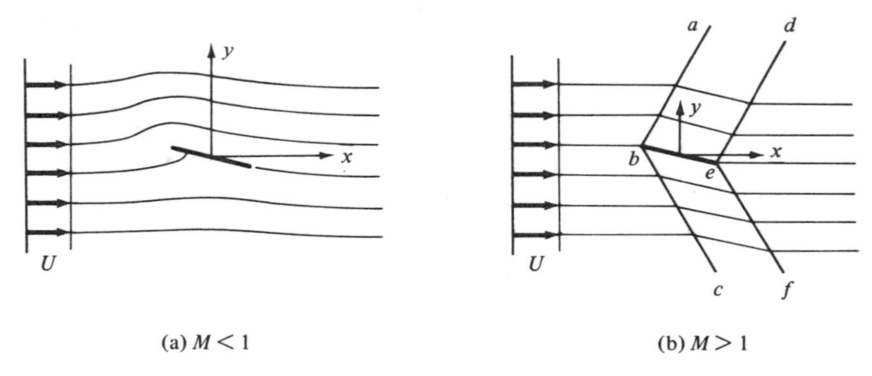

(a) $M < 1$ (b) $M > 1$

FIGURE 1.2 Subsonic and supersonic flow past flat plate airfoil.

The case where $M \approx 1$, the so-called "transonic" regime, is especially interesting—and difficult. The approximate equation (1.3) is then inaccurate, and a nonlinear term, neglected in deriving (1.3), must be retained. By means of a complicated change of both dependent and independent variables, Chaplygin[4] was able to suppress the nonlinearity, and arrived at what is now called the Tricomi equation

[1] That is, the x and y fluid velocity components are given by u_x and u_y, respectively. (Actually, the velocity potential is generally denoted by the letter ϕ. However, we choose to always use u for the unknown, in order to minimize the nomenclature and, we hope, to make for easier reading.)

[2] If, in fact, the fluid is considered to be incompressible, then the speed of sound is infinite, so that $M = 0$, and (1.3) reduces to the well-known Laplace equation.

[3] Specifically, u is an analytic function of x and y throughout the flow.

[4] See, for example, L.D. Landau and E.M. Lifshitz, *Fluid Mechanics*, Addison-Wesley Publishing Company, Reading, Mass., 1959, Chapter XII.

$$u_{xx} - xu_{yy} = 0 \tag{1.4}$$

where the "hodograph variables" x, y, u are different from, but related to, the "physical variables" in (1.3). Now, $B^2 - AC = x$, so that (1.4) is elliptic in the half-plane $x < 0$, and hyperbolic in the half-plane $x > 0$! This reflects the fact that the flow is mixed in the physical x, y plane; subsonic in some region(s) and supersonic elsewhere.

As in PART I, our development will be based upon a detailed discussion of several representative examples. In the remainder of this section we will introduce the equations which will be considered in Sections 2 through 10. In doing so, we will assign a definite physical setting to each equation; first, so that we may be guided by whatever physical intuition is available to us and, second, so that our results may be interpreted and understood more fully. Whereas each equation is associated with a wide range of physical phenomena, we have selected only a few—which should be reasonably familiar to most readers.

Elliptic. As representative of the elliptic type, we will consider the classical **Poisson** and **Helmholtz** equations,[5]

$$\mathbf{L}u = \nabla^2 u = \phi \tag{1.5}$$

$$\mathbf{L}u = \nabla^2 u + k^2 u = \phi \tag{1.6}$$

respectively. Of the various physical phenomena governed by the Poisson equation, we will consider it in connection with the *gravitational potential.* Specifically, equation (1.5) governs the gravitational potential $u(x, y)$ induced by a mass distribution of density $\phi(x, y)$ thoughout the region under consideration; the resulting force field is given by $-\nabla u$.

As for an interpretation of the Helmholtz equation (1.6), let us consider first the forced *vibration of a stretched membrane*, such as a drumhead. The deflection $w(x, y, t)$, measured positive "upward," is governed by the two-dimensional wave equation

$$c^2 \nabla^2 w - w_{tt} = fe^{i\omega t} \tag{1.7}$$

where the constant c^2 is the tension per unit length divided by the mass per unit area, and $f(x, y)$ is an applied force per unit area divided by the mass per unit area, measured positive "downward." Since the driving force is harmonic,

[5] The two-dimensional *Laplacian* differential operator, ∇^2 (read "del square," and denoted as Δ by some authors), denotes the vector dot product $\nabla \cdot \nabla$, where ∇ is the *gradient* operator, defined in Cartesian coordinates as $\nabla \equiv \mathbf{i}\partial/\partial x + \mathbf{j}\partial/\partial y$. Thus,

$$\nabla^2 = \nabla \cdot \nabla = \left(\mathbf{i}\frac{\partial}{\partial x} + \mathbf{j}\frac{\partial}{\partial y}\right) \cdot \left(\mathbf{i}\frac{\partial}{\partial x} + \mathbf{j}\frac{\partial}{\partial y}\right) = \frac{\partial^2}{\partial x^2} + \frac{\partial^2}{\partial y^2}$$

since $\mathbf{i}, \mathbf{j}$ are unit vectors (vectors will be denoted by boldface type) in the x, y directions, respectively, with $\mathbf{i} \cdot \mathbf{j} = \mathbf{j} \cdot \mathbf{i} = 0$ and $\mathbf{i} \cdot \mathbf{i} = \mathbf{j} \cdot \mathbf{j} = 1$.

it is reasonable to seek $w(x, y, t)$ in the form $u(x, y) \exp(i\omega t)$. Inserting this into (1.7) and canceling the exponential, we obtain the Helmholtz equation (1.6) on $u(x, y)$, with $k^2 = \omega^2/c^2$ and $\phi(x, y) = f(x, y)/c^2$. Basically, then, we may regard k and u in (1.6) as the frequency and amplitude of the displacement of a membrane which is driven by a periodic force distribution of amplitude ϕ.

The Helmholtz equation will also be discussed in connection with the propagation of weak two-dimensional disturbances in still air, i.e., two-dimensional *acoustics*. Here, the governing equation is

$$c^2\nabla^2\psi = \psi_{tt}; \qquad \nabla^2 = \frac{\partial^2}{\partial x^2} + \frac{\partial^2}{\partial y^2} \tag{1.8}$$

where ψ is the velocity potential (i.e., the air velocity field is given by $\nabla\psi$) and c is the speed of sound. Again, seeking $\psi(x, y, t)$ in the form $u(x, y) \exp(i\omega t)$ leads to a Helmholtz equation on $u(x, y)$.

Parabolic. For the parabolic type, we will consider the classical one-dimensional **diffusion** equation

$$Lu = \kappa u_t - u_{xx} = \phi \tag{1.9}$$

within the context of *heat conduction*. Thus, $u(x, t)$ is the temperature, t is the time, κ is a physical constant which is a measure of the thermal conductivity, and $\phi(x, t)$ is the applied heat source density per unit x-length.

Hyperbolic. Finally, as representative of the hyperbolic type we will consider the one-dimensional **wave** equation

$$Lu = c^2u_{xx} - u_{tt} = \phi \tag{1.10}$$

in connection with the *lateral motion of a taut string*. Thus, $u(x, t)$ is the displacement, measured positive "upward," t is the time, c^2 is the tension divided by the mass per unit length, and $\phi(x, t)$ is an applied lateral force per unit length, measured positive "downward," divided by the tension.

EXERCISES

1.1. Classify the following partial differential equations, as to whether they are elliptic, parabolic, or hyperbolic. If they are of mixed type, specify the various regions and corresponding classifications.

(a) $u_{xx} = p(x)u_{tt} + h(x)u_t + f(x)e^{i\omega t}; \qquad p(x) > 0$

(b) $u_{xx} + (1 - x)^2u_{yy} = 6$

(c) $u_{xy} + xyu_{yy} - u_x = 0$

(d) $u_{xx} = p(x)u_t + h(x)u + f(x, t)$

2. THE ADJOINT OPERATOR

We introduce the **formal adjoint L***, associated with **L**, in the same way that we did for ordinary differential operators in PART I, using integration by parts:

$$\iint_{\mathscr{S}} v\mathbf{L}u\, d\sigma = \text{boundary terms} + \iint_{\mathscr{S}} u\mathbf{L}^* v\, d\sigma \tag{2.1}$$

where $d\sigma$ is a differential element of area of the region $\mathscr{S}$ under consideration; c.f. the analogous equation (2.1) of PART I. In the present case, however, the boundary terms will be in the form of a line integral over $\mathscr{C}$, the boundary of $\mathscr{S}$.

To illustrate the procedure, let us limit our attention to the first term of **L**u, from (1.1), namely Au_{xx}. Referring to Fig. 2.1, we integrate by parts as follows,

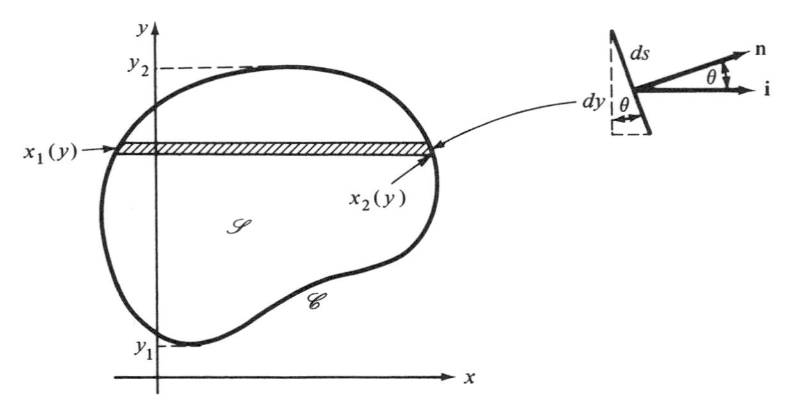

FIGURE 2.1 Integration by parts.

$$\begin{aligned}
\iint_{\mathscr{S}} vAu_{xx}\, d\sigma &= \int_{y_1}^{y_2}\left\{\int_{x_1(y)}^{x_2(y)} vAu_{xx}\, dx\right\} dy \\
&= \int_{y_1}^{y_2}\left\{vAu_x\Big|_{x_1(y)}^{x_2(y)} - \int_{x_1(y)}^{x_2(y)} (vA)_x u_x\, dx\right\} dy \\
&= \int_{y_1}^{y_2}\left\{[vAu_x - (vA)_x u]\Big|_{x_1(y)}^{x_2(y)} + \int_{x_1(y)}^{x_2(y)} (vA)_{xx} u\, dx\right\} dy \qquad (2.2)\\
&= \int_{y_1}^{y_2} [vAu_x - (vA)_x u]\Big|_{x_1(y)}^{x_2(y)} dy + \iint_{\mathscr{S}} (vA)_{xx} u\, d\sigma \\
&= \int_{\mathscr{C}} [vAu_x - (vA)_x u]\mathbf{i}\cdot\mathbf{n}\, ds + \iint_{\mathscr{S}} (vA)_{xx} u\, d\sigma
\end{aligned}$$

since (Fig. 2.1) $dy = ds \cos\theta = \mathbf{n} \cdot \mathbf{i}\, ds$, where $\mathbf{n}$ is the outward unit normal, and ds is a differential element of arc-length along $\mathscr{C}$.[6]

Treating each term in $v\mathbf{L}u$ in this manner (Exercise 2.1), we obtain the desired form

$$\iint_{\mathscr{S}} v\mathbf{L}u \, d\sigma = \int_{\mathscr{C}} (M\mathbf{i} + N\mathbf{j}) \cdot \mathbf{n}\, ds + \iint_{\mathscr{S}} u\mathbf{L}^* v \, d\sigma \tag{2.3}$$

where

$$\begin{aligned} \mathbf{L}^* v &= (Av)_{xx} + 2(Bv)_{xy} + (Cv)_{yy} - (Dv)_x - (Ev)_y + Fv \\ M &= Avu_x - u(Av)_x + 2vBu_y + Duv \\ N &= -2u(Bv)_x + Cvu_y - u(Cv)_y + Euv \end{aligned} \tag{2.4}$$

Equation (2.3) is important. It will serve as the starting point for our Green's function approach, in the same way that the analogous formula did for ordinary differential equations in PART I. It is worth emphasizing that the arc-length increments ds in these formulas are, by their definition, always *positive*. The relevance of this remark will be made clear later on.

Let us rewrite this result explicitly for the four differential operators defined above in (1.5), (1.6), (1.9), and (1.10). For the Laplace differential operator

$$\mathbf{L}u = \nabla^2 u \tag{2.5}$$

we have $A = C = 1$ and $B = D = E = F = 0$, so that $M = vu_x - uv_x$ and $N = vu_y - uv_y$. Noting that

$$(M\mathbf{i} + N\mathbf{j}) \cdot \mathbf{n} = (v\nabla u - u\nabla v) \cdot \mathbf{n} = vu_n - uv_n \tag{2.6}$$

where $u_n \equiv \partial u/\partial n$ and $v_n \equiv \partial v/\partial n$, (2.3) becomes

$$\iint_{\mathscr{S}} v\mathbf{L}u \, d\sigma = \int_{\mathscr{C}} (vu_n - uv_n)\, ds + \iint_{\mathscr{S}} u\mathbf{L}^* v \, d\sigma \tag{2.7}$$

where

$$\mathbf{L}^* v = \nabla^2 v \tag{2.8}$$

For the Helmholtz differential operator $\mathbf{L}u = \nabla^2 u + k^2 u$ we find that equation (2.7) again applies, $\mathbf{L}^* v = \nabla^2 v + k^2 v$. For the diffusion and wave

[6] Note that we have tacitly assumed, in this calculation, that the region $\mathscr{S}$ is *convex* in the x direction; i.e., each cross-hatched "sliver" running from $x_1(y)$ to $x_2(y)$ lies entirely within $\mathscr{S}$. There is little point in spelling out the details for the nonconvex case, since the idea is basically the same, and the final result [the last line of (2.2)] remains unchanged.

Actually, our derivation of (2.2), which appeals only to fundamentals, is equivalent to the application of the *divergence theorem*, with which the reader is very likely familiar.

differential operators, $\mathbf{L}u = \kappa u_t - u_{xx}$ and $\mathbf{L}u = c^2 u_{xx} - u_{tt}$ respectively, (2.3) becomes

$$\iint_{\mathscr{S}} v\mathbf{L}u\, d\sigma = \int_{\mathscr{C}} [(uv_x - vu_x)\mathbf{i} + \kappa uv\mathbf{j}] \cdot \mathbf{n}\, ds + \iint_{\mathscr{S}} u\mathbf{L}^* v\, d\sigma \tag{2.9}$$

and

$$\iint_{\mathscr{S}} v\mathbf{L}u\, d\sigma = \int_{\mathscr{C}} [c^2(vu_x - uv_x)\mathbf{i} - (vu_t - uv_t)\mathbf{j}] \cdot \mathbf{n}\, ds + \iint_{\mathscr{S}} u\mathbf{L}^* v\, d\sigma \tag{2.10}$$

where $\mathbf{L}^* v = -\kappa v_t - v_{xx}$ and $\mathbf{L}^* v = c^2 v_{xx} - v_{tt}$, respectively.

Of the four differential operators under consideration, all are seen to be formally self-adjoint, except for the diffusion differential operator. It can, in fact, be shown (Exercise 2.3) that the relations

$$\begin{aligned} A_x + B_y &= D \\ B_x + C_y &= E \end{aligned} \tag{2.11}$$

constitute necessary and sufficient conditions for the formal self-adjointness of the second order differential operator defined in (1.1).

As in PART I, it is convenient to distinguish between the **differential operator L,** and the **operator $\mathscr{L}$** consisting of **L** plus the imposed boundary conditions. For the case where the boundary conditions are homogeneous, we define not only a **formal adjoint differential operator L***, but also an **adjoint operator $\mathscr{L}^*$**. As before, $\mathscr{L}^*$ is defined by the relation

$$(\mathbf{L}u, v) = (u, \mathbf{L}^* v) \tag{2.12}$$

where the two-dimensional *inner product* of two functions, say f and g, is defined here by

$$(f, g) = \iint_{\mathscr{S}} fg\, d\sigma \tag{2.13}$$

Comparing (2.12) with (2.3), we observe that $\mathscr{L}^*$ consists of the differential operator $\mathbf{L}^*$ plus homogeneous boundary conditions which are such that the boundary terms, arising through the integration by parts, all vanish.

For example, consider $\mathscr{L}$ to consist of $\mathbf{L} = \nabla^2$ over the disk $x^2 + y^2 < 1$, together with the boundary condition $u = 0$ on the circle $x^2 + y^2 = 1$. We've already seen, from (2.8), that $\mathbf{L}$ is formally self-adjoint. Now examine the boundary term in the integration by parts formula (2.7). Whereas $u = 0$ on $\mathscr{C}$, the normal derivative u_n is not specified, and hence need not be zero. For the boundary terms to vanish we therefore need $v = 0$ on $\mathscr{C}$. Since the adjoint boundary condition, $v = 0$ on $\mathscr{C}$, is identical to the original boundary condition, $u = 0$ on $\mathscr{C}$, and $\mathbf{L}^* = \mathbf{L}$ as well, it follows that $\mathscr{L}^* = \mathscr{L}$, i.e., $\mathscr{L}$ is **self-adjoint.**

As one more example, consider $\mathscr{L}$ to consist of the wave differential operator $\mathbf{L} = c^2\partial^2/\partial x^2 - \partial^2/\partial t^2$ over the rectangular region $0 < x < 1$, $0 < t < T$ in the x, t plane, plus the following homogeneous boundary conditions: end conditions $u(0, t) = u(1, t) = 0$ and initial conditions $u(x, 0) = u_t(x, 0) = 0$, as indicated in Fig. 2.2. Having already seen that $\mathbf{L}^* = \mathbf{L}$ for the wave differential operator, let us determine the adjoint boundary conditions. Breaking up $\mathscr{C}$ into its four straight sides, the boundary term in (2.10) becomes

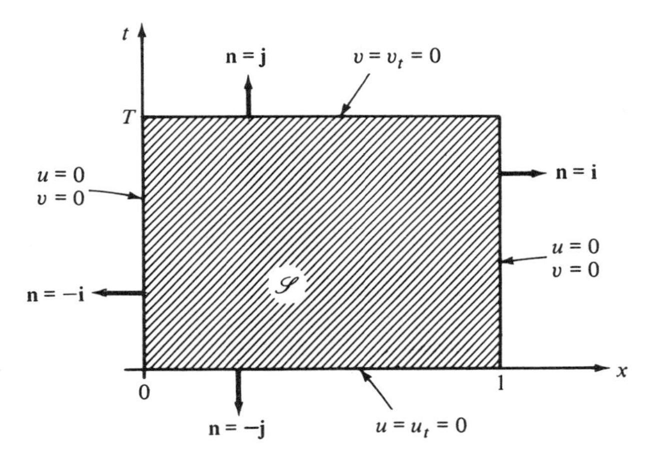

FIGURE 2.2 Adjoint boundary conditions for wave operator.

$$\int_{\mathscr{C}} [c^2(vu_x - uv_x)\mathbf{i} - (vu_t - uv_t)\mathbf{j}] \cdot \mathbf{n}\, ds$$
$$= -c^2 \int_0^T (vu_x - uv_x)\Big|_{x=0} dt - \int_0^1 (vu_t - uv_t)\Big|_{t=T} dx \qquad (2.14)$$
$$+ c^2 \int_0^T (vu_x - uv_x)\Big|_{x=1} dt + \int_0^1 (vu_t - uv_t)\Big|_{t=0} dx$$

[Recall our earlier remark that the ds increments are to be *positive*. With this in mind, be sure to understand the signs and limits of the four integrals on the right-hand side of (2.14).] Noting our boundary conditions on u, we see that the adjoint boundary conditions must be $v(0, t) = v(1, t) = 0$ and $v(x, T) = v_t(x, T) = 0$, if the boundary terms (2.14) are to vanish. These are *not* identical to the original boundary conditions (Fig. 2.2),[7] so that $\mathscr{L}$ is not self-adjoint, even though $\mathbf{L}$ is formally self-adjoint.

[7] They *would* be if we had $v(x, 0) = v_t(x, 0) = 0$ instead of $v(x, T) = v_t(x, T) = 0$.

EXERCISES

2.1. Derive the integration by parts formulas (2.3) and (2.4) using the procedure outlined in equation (2.2).

2.2. Show that M and N, as given by (2.4), are not uniquely determined. *Hint:* We can add Ψ_y to M and $-\Psi_x$ to N, where $\Psi(x, y)$ is arbitrary.

2.3. Verify that the relations (2.11) do, in fact, constitute necessary and sufficient conditions for the formal self-adjointness of $\mathbf{L}$. *Hint:* To prove that they are necessary, set $u = 1$, x, and y, in turn.

2.4. Determine the adjoint operator $\mathscr{L}^*$ corresponding to each of the following operators:
(a) $\mathbf{L} = \nabla^2$ over $x^2 + y^2 < 1$; $\alpha u + \beta u_n = 0$ on $x^2 + y^2 = 1$.
(b) $\mathbf{L} = \kappa\partial/\partial t - \partial^2/\partial x^2$ over the rectangle $0 < x < 1$, $0 < t < T$; $u(0, t) = u(1, t) = u(x, 0) = 0$. Are these operators self-adjoint?

3. THE DELTA FUNCTION

Having already discussed the one-dimensional delta function $\delta(x - \xi)$ in some detail, we will extend the definition to two dimensions with a minimum of discussion. Specifically, we define $\delta(x - \xi, y - \eta)$ such that

$$\iint_{\mathscr{S}} \delta(x - \xi, y - \eta)h(x, y)\, d\sigma = h(\xi, \eta) \tag{3.1}$$

for every h which is continuous over the region $\mathscr{S}$, which contains the point ξ, η.

As in the case of the one-dimensional delta function $\delta(x - \xi)$, we note that $\delta(x - \xi, y - \eta)$ may be visualized as the formal limit of a sequence of ordinary functions. Symbolically,

$$\delta(x - \xi, y - \eta) = \lim_{k\to\infty} w_k(r) \tag{3.2}$$

where

$$r = \sqrt{(x - \xi)^2 + (y - \eta)^2} \tag{3.3}$$

For example,

$$w_k(r) = \begin{cases} \dfrac{k^2}{\pi}, & r < \dfrac{1}{k} \\[2ex] 0, & r \geq \dfrac{1}{k} \end{cases} \tag{3.4}$$

and

$$w_k(r) = \frac{ke^{-kr^2}}{\pi} \tag{3.5}$$

are two-dimensional *δ-sequences.*

Finally, we note that

$$\iint_{\mathscr{S}} \delta(x - \xi)\delta(y - \eta)h(x, y)\, dx\, dy = h(\xi, \eta) \tag{3.6}$$

Comparing (3.1) and (3.6), we see that

$$\delta(x - \xi, y - \eta) = \delta(x - \xi)\delta(y - \eta) \tag{3.7}$$

so that the two-dimensional delta function can be expressed as the product of one-dimensional delta functions.

EXERCISES

3.1. Verify that the δ-sequence (3.4) satisfies the fundamental property

$$\lim_{k \to \infty} \int_{-\infty}^{\infty}\int_{-\infty}^{\infty} w_k(r)h(x, y)\, dx\, dy = h(\xi, \eta)$$

[This is the two-dimensional analog of equation (3.10) of PART I, generalized so that the delta function acts at ξ, η instead of at the origin.]

4. THE GREEN'S FUNCTION METHOD

Our development of the Green's function method for the partial differential equation (1.1) will follow essentially the same lines as for the case of ordinary differential equations in PART I. Our starting point is equation (2.3), or its more specific form, such as equation (2.7), (2.9), or (2.10).

Briefly, we choose "v" to be our Green's function G, and use x, y to denote the fixed point, with ξ, η as the corresponding dummy integration variables. Requiring that G satisfy

$$\mathbf{L}^* G = \delta(\xi - x, \eta - y) \tag{4.1}$$

over $\mathscr{S}$, the last term in (2.3) reduces to $u(x, y)$. Replacing $\mathbf{L}u$ in the left-hand side by the known function $\phi(\xi, \eta)$, and subjecting G to boundary conditions on $\mathscr{C}$ which result in the elimination of any "unwelcome" terms in the boundary integral, we are able to solve for the desired $u(x, y)$ in terms of quantities which are known—provided that we are, in fact, able to determine the Green's function.

The whole point of the method is that the boundary value problem governing G is generally somewhat simpler than the original one governing u. Sometimes, as we shall see, G can in fact be constructed simply by inspection.

In general, it is convenient to seek G in two parts,

$$G(\xi, \eta; x, y) = U(\xi, \eta; x, y) + g(\xi, \eta; x, y) \tag{4.2}$$

where U is a particular solution of (4.1), which need not satisfy the required boundary conditions, and g is a solution of the homogeneous equation $\mathbf{L}^* G = 0$, such that the combination $U + g$ does satisfy those boundary conditions. We will refer to U as a **principal solution**, although it is also called a *fundamental solution*, an *elementary solution*, and a *free-space Green's function* in the literature. It contains the basic singularity of the Green's function. The hard part, in constructing the Green's function, is generally in the determination of the "regular" part g so as to satisfy the required boundary conditions.

To illustrate this "splitting" technique, which was not employed in PART I (but could have been), let us use it to find the Green's function for the simple loaded string example in PART I (Example 1, Section 4). The conditions required of G were:

$$\mathbf{L}^* G = G_{\xi\xi} = \delta(\xi - x) \tag{4.3a}$$

$$G(0, x) = G(1, x) = 0 \tag{4.3b}$$

We seek $G = U + g$, where U satisfies

$$\mathbf{L}^* U = U_{\xi\xi} = \delta(\xi - x) \tag{4.4}$$

subject to no particular boundary conditions. Integrating (4.4) twice,

$$\begin{aligned} U_\xi &= H(\xi - x) \\ U &= (\xi - x)H(\xi - x) \end{aligned} \tag{4.5}$$

This does contain the basic singularity, namely the "kink" at $\xi = x$, but does not satisfy the boundary conditions (4.3b) since $U(0, x) = 0$, $U(1, x) = 1 - x$. Now, g is to satisfy the homogeneous equation

$$\mathbf{L}^* g = g_{\xi\xi} = 0 \tag{4.6}$$

subject to boundary conditions which are such that the combination $U + g$ satisfies (4.3b). Well,

$$\begin{aligned} G(0, x) = 0 &= U(0, x) + g(0, x) \\ &= 0 + g(0, x) \end{aligned} \tag{4.7a}$$

$$\begin{aligned} G(1, x) = 0 &= U(1, x) + g(1, x) \\ &= 1 - x + g(1, x) \end{aligned} \tag{4.7b}$$

so that

$$g(0, x) = 0, \qquad g(1, x) = x - 1 \tag{4.8}$$

Solving (4.6),

$$g(\xi, x) = A\xi + B \tag{4.9}$$

and, invoking the boundary conditions (4.8) to determine A and B, we obtain

$$g(\xi, x) = (x - 1)\xi \tag{4.10}$$

Finally,

$$G(\xi, x) = U + g = (\xi - x)H(\xi - x) + (x - 1)\xi \tag{4.11}$$

Of course, this was an extremely simple example, but it does serve to illustrate the "splitting" technique that we will be using.

Before getting into applications, let us first consider the determination of the principal solutions for the particular partial differential operators under consideration here.

EXERCISES

4.1. Plot U, g and $G = U + g$, as functions of ξ, for the preceding example.
4.2. How are linearity and superposition involved in the splitting technique?

5. PRINCIPAL SOLUTIONS

Let us now determine principal solutions for the Laplace, Helmholtz, diffusion, and wave differential operators.

Laplace Operator. Recalling (4.1), we require that the principal solution U satisfy

$$\mathbf{L}^* U = \nabla^2 U = \delta(\xi - x, \eta - y) \tag{5.1}$$

It will prove helpful, in the following, to have a physical interpretation of the principal solution U. In line with our gravitational interpretation of the Poisson equation (1.5), we see that U may, in fact, be regarded as the potential induced at a field point ξ, η by a point mass of unit strength at x, y.[8] It should therefore be symmetric about the point x, y and will depend only on the radial variable

[8] More precisely, a line distribution of unit mass per unit length, located at x, y and extending from $-\infty$ to $+\infty$ in the normal z direction (Exercise 5.1). For brevity, we shall refer to it as a "point mass of unit strength," with this obvious meaning.

$$r = \sqrt{(\xi - x)^2 + (\eta - y)^2} \tag{5.2}$$

Replacing the delta function in (5.1) by the δ-sequence (3.5), with the understanding that we will let $k \to \infty$ in the final expression for U, and expressing the Laplacian in terms of polar coordinates r, θ centered at the x, y point, (5.1) becomes

$$\nabla^2 U = \frac{1}{r}(rU_r)_r + \frac{1}{r^2}U_{\theta\theta} = \frac{ke^{-kr^2}}{\pi} \tag{5.3}$$

where $U_{\theta\theta} = 0$ since U depends only on r. Multiplying through by r, and integrating from 0 to r, we obtain

$$rU_r = \frac{1 - e^{-kr^2}}{2\pi} \tag{5.4}$$

where we have imposed the condition that $U_r(0)$ be finite. Requiring further that $U(1) = 0$, for definiteness, we may integrate (5.4) from r to 1, and hence obtain

$$U(r) = \frac{1}{2\pi}\ln r + \frac{1}{2\pi}\int_r^1 \frac{e^{-kt^2}}{t}\,dt \tag{5.5}$$

For each fixed $r > 0$, the integral term tends to zero as $k \to \infty$, since $\exp(-kt^2) \to 0$ *uniformly* over $r \leq t \leq 1$ as $k \to \infty$, so that we have as our principal solution (see Exercise 5.8)

$$U = \frac{1}{2\pi}\ln r \tag{5.6}$$

Alternative Approach. An approach which is more direct, though somewhat formal, is as follows. For $r > 0$, so that we are away from the delta function singularity, we have from (5.1)

$$\nabla^2 U = \frac{1}{r}(rU_r)_r = 0 \tag{5.7}$$

Integrating twice we easily obtain

$$U = A \ln r + B \tag{5.8}$$

Thus far we have considered $r > 0$, so that the delta function behavior has not yet been taken into account. With this in mind, let us seek A and B by integrating the governing equation (5.1) over a disk of arbitrarily small radius ϵ, centered at x, y.

$$\iint \nabla^2 U d\sigma = \iint \delta \, d\sigma \tag{5.9}$$

Applying the Gauss divergence theorem,[9] and using (5.8), the left-hand side becomes

$$\int U_r\Big|_{r=\epsilon} ds = \frac{A}{\epsilon}\int ds = 2\pi A \tag{5.10}$$

Since the right-hand side of (5.9) is unity, it follows that $A = 1/2\pi$; B remains arbitrary, and can therefore be set equal to zero for convenience. Our result is then in agreement with (5.6). Of course, application of the divergence theorem is rather formal since U is not suitably well behaved over the disk.

Helmholtz Operator. We require that U satisfy

$$\mathbf{L}^* U = \nabla^2 U + k^2 U = \delta(\xi - x, \eta - y) \tag{5.11}$$

Introducing polar coordinates at the point x, y as before, and seeking U as a function of r alone, we have for $r > 0$

$$\frac{1}{r}(rU_r)_r + k^2 U = 0 \tag{5.12}$$

or

$$rU_{rr} + U_r + k^2 rU = 0 \tag{5.13}$$

The change of variables $\rho = kr$ reduces this to a Bessel equation[10] of order zero, with linearly independent solutions $J_0(\rho)$ and $Y_0(\rho)$ or $J_0(kr)$ and $Y_0(kr)$. Of these two, the *Bessel function of first kind and order zero*, J_0, can be discarded since it is regular at $r = 0$ and hence cannot lead to the desired delta function singularity. We have high hopes for the *Bessel function of second kind and order zero*, however, due to its singular behavior

$$Y_0(kr) \sim \left(\frac{2}{\pi}\right) \ln r \tag{5.14}$$

as $r \to 0$, and so we tentatively set

$$U = AY_0(kr) \tag{5.15}$$

As above, we attempt to determine the scale factor A by integrating the

[9] That is, we apply it to a cylinder of unit height, with our ϵ-disk as its base. Since U is independent of the normal variable, say z, we have

$$\iint \nabla^2 U\, d\sigma = \iiint \nabla^2 U\, d\tau = \iint \nabla U \cdot \mathbf{n}\, d\sigma = \iint \frac{\partial U}{\partial n}\, d\sigma$$

where the second equality follows from Gauss's theorem. On the top and bottom faces of the cylinder $\partial U/\partial n = 0$. On the side, $\partial U/\partial n = \partial U/\partial r$, and $d\sigma = 1\, ds$.

[10] Recall that some information on Bessel functions was given in PART I (Example 4 of Section 4).

governing equation (5.11) over an arbitrarily small disk of radius ϵ, centered at x, y. The only difference is the k^2U term, present in (5.11) but not in (5.1). When (5.11) is integrated over the disk, however, the effect of this term is nil, since its integral is of order $O(\epsilon^2 \ln \epsilon)$ (the ϵ^2 from the area of the disk, and the $\ln \epsilon$ from the small r behavior of U), which tends to zero[11] as $\epsilon \to 0$. Applying the divergence theorem to the $\iint \nabla^2 U \, d\sigma$ term, as before, and noting from (5.14) and (5.15) that $U \sim (2A/\pi) \ln r$, we find that $A = \frac{1}{4}$, so that

$$U = \frac{1}{4} Y_0(kr) \tag{5.16}$$

Diffusion Operator. We require that

$$\mathbf{L}^* U = -\kappa U_\tau - U_{\xi\xi} = \delta(\xi - x, \tau - t) \tag{5.17}$$

for $0 < \tau < \infty$ and $-\infty < \xi < \infty$. In the case of the Laplace and Helmholtz differential operators treated above, the two independent variables ξ and η appeared *symmetrically* in the partial differential equations. This implied that U could be considered to be symmetric about the singular point x, y; it therefore depended only on the radial variable r, so that the equations reduced to ordinary differential equations which were easily solved. Unfortunately, no such symmetry is available for the diffusion operator, and the determination of U will, accordingly, be somewhat more difficult.

Before starting, let us try to anticipate the nature of U based upon a physical interpretation of the situation. To do so, it is convenient to consider first the situation that would exist if U were governed instead by the equation

$$\mathbf{L} U = \kappa U_\tau - U_{\xi\xi} = \delta(\xi - x, \tau - t) \tag{5.18}$$

since this has already been interpreted in Section 1 in terms of one-dimensional heat conduction. Specifically, U would then be identically zero for all times $\tau < t$. At $\tau = t$ a unit heat source would be applied instantaneously at $\xi = x$, and U would describe the subsequent diffusion of that heat pulse for times $\tau > t$. Now, returning to (5.17) we see that the situation is identical, except that *time is reversed*. Thus, our principal solution will be zero for all $\tau > t$. A

[11] More generally, consider $\epsilon^\alpha \ln \epsilon$ as $\epsilon \to 0$, where $\alpha > 0$. This is indeterminate since $\epsilon^\alpha \to 0$ while $\ln \epsilon \to -\infty$. However, rewriting it as the fraction $\ln \epsilon / \epsilon^{-\alpha}$ and applying L'Hospital's rule, we find that the limit does exist as $\epsilon \to 0$; specifically, $\epsilon^\alpha \ln \epsilon \to 0$ for any $\alpha > 0$. This is very interesting because it exposes $\ln \epsilon$ as being a *very weak* singularity; weaker than $\epsilon^{-\alpha}$, where α is an *arbitrarily small* positive number! (Similarly, L'Hospital's rule shows that $\ln x$ has an equally weak singularity at *infinity*, i.e., $\ln x / x^\alpha \to 0$ as $x \to \infty$ for any $\alpha > 0$. In a sense, the singularities of the logarithm, at zero and infinity, are "infinitely weak.")

heat pulse is applied at $\tau = t$, and U describes the diffusion of this pulse as τ *decreases*. Whether or not this situation is possible, physically, is beside the point since (5.17) is only a formal mathematical definition of U; the *physics* is contained in the original equation (1.5).

To solve (5.17) we choose to Fourier transform on ξ. This produces the ordinary differential equation

$$-\kappa \hat{U}_\tau + \omega^2 \hat{U} = \delta(\tau - t)e^{-i\omega x} \tag{5.19}$$

where ω is the transform variable. Noting that the right-hand side is zero for $\tau > t$ and $\tau < t$, we obtain

$$\hat{U} = \begin{cases} Ae^{\omega^2\tau/\kappa}, & \tau > t \\ Be^{\omega^2\tau/\kappa}, & \tau < t \end{cases} \tag{5.20}$$

We blend these two halves of the solution by integrating (5.19) on τ, from $t - 0$ to $t + 0$. This implies the following relationship between A and B,

$$-\kappa Ae^{\omega^2 t/\kappa} + \kappa Be^{\omega^2 t/\kappa} = e^{-i\omega x} \tag{5.21}$$

As a second relation we recall that U, and hence $\hat{U}$, is identically zero for $\tau > t$. Thus $A = 0$, and

$$\hat{U} = \kappa^{-1} H(t - \tau)e^{-i\omega x - \omega^2(t-\tau)/\kappa} \tag{5.22}$$

Inverting (see Exercise 5.2), we obtain the result

$$\begin{aligned} U &= \frac{H(t-\tau)}{2\pi\kappa} \int_{-\infty}^{\infty} e^{i\omega(\xi - x) - \omega^2(t-\tau)/\kappa}\, d\omega \\ &= \frac{H(t-\tau)e^{-\kappa(\xi - x)^2/4(t-\tau)}}{\sqrt{4\pi\kappa(t-\tau)}} \end{aligned} \tag{5.23}$$

Wave Operator. In this case we require

$$\mathbf{L}^* U = c^2 U_{\xi\xi} - U_{\tau\tau} = \delta(\xi - x, \tau - t) \tag{5.24}$$

for $0 < \tau < \infty$ and $-\infty < \xi < \infty$. That is, U is identically zero for all $\tau < t$; at $\tau = t$ a unit impulsive force is applied to the string instantaneously at $\xi = x$, and U describes its subsequent displacement. We leave it as an exercise (Exercise 5.3) to show that

$$U = \begin{cases} -1/2c, & |\xi - x| < c(\tau - t) \\ 0, & |\xi - x| > c(\tau - t) \end{cases} \tag{5.25}$$

as indicated in Fig. 5.1. The wedge-shaped region, with apex at x, t, is generally known as the *range of influence* of the point x, t since it is only within this part of the ξ, τ plane that the effect of the pulse at x, t is felt.

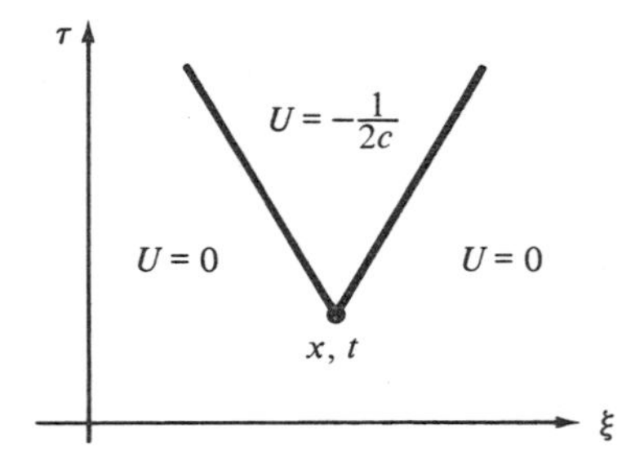

FIGURE 5.1 Principal solution of wave equation.

Now, observe that if some function, say $f(\xi - x, \tau - t)$, is a solution of (5.24) then so is $f(\xi - x, t - \tau)$. To see this, introduce the change of variables $\tau - t = t - \tau'$. Then, since $f(\xi - x, \tau - t)$ satisfies (5.24),

$$c^2 f_{\xi\xi}(\xi - x, t - \tau') - f_{\tau\tau}(\xi - x, t - \tau') = \delta(\xi - x, t - \tau')$$

But $f_{\tau\tau}(\xi - x, t - \tau') = f_{\tau'\tau'}(\xi - x, t - \tau')$ since $\tau = -\tau' + 2t$, and $\delta(\xi - x, t - \tau') = \delta(\xi - x, \tau' - t)$, so that

$$c^2 f_{\xi\xi}(\xi - x, t - \tau') - f_{\tau'\tau'}(\xi - x, t - \tau') = \delta(\xi - x, \tau' - t)$$

Surely this equation must remain true if we replace τ' by some other letter; replacing it by τ, we see that $f(\xi - x, t - \tau)$ does in fact satisfy (5.24), as claimed.

Thus, replacing $\tau - t$ in (5.25) by $t - \tau$ we obtain another principal solution,

$$U = \begin{cases} -1/2c, & |\xi - x| < c(t - \tau) \\ 0, & |\xi - x| > c(t - \tau) \end{cases} \tag{5.26}$$

as indicated in Fig. 5.2. It is identical to (5.25), except that it propagates "backwards."

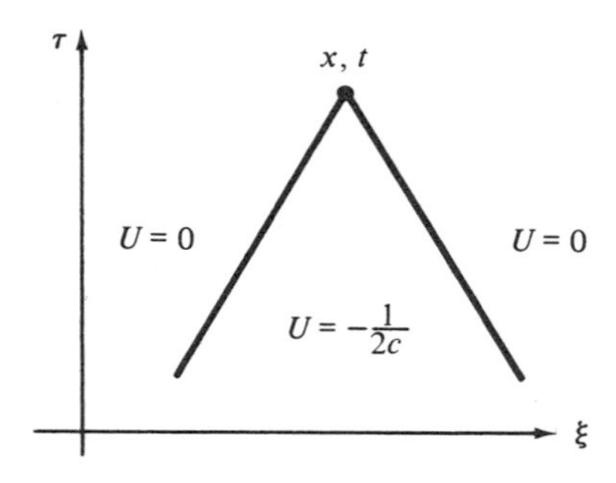

FIGURE 5.2 "Backward running" principal solution.

In this case, the wedge-shaped region is called the *domain of dependence* since x, t is within the range of influence of all points within the wedge. In other words, any disturbance created within the wedge will be felt at x, t. These ideas come under the heading of what is generally known as *causality*.

For obvious reasons, we shall subsequently refer to (5.25) and (5.26) as the **forward running** and **backward running principal solutions** (the terms *advancing* and *retarded* are also used by some authors).

EXERCISES

5.1. Based upon our physical interpretation of the Poisson equation (1.5), and the form of (5.1), we stated that $U = (1/2\pi) \ln r$ can therefore be interpreted as the gravitational potential induced by a point mass of unit strength at x, y (more precisely, a straight line distribution of mass, of unit strength per unit length, located at x, y and extending from $-\infty$ to $+\infty$ in the normal z direction). Verify this claim, starting with the law of gravitational attraction; namely, the gravitational force per unit mass induced by a point mass m at another point is $m/4\pi\, d^2$ where d is the separation distance. Thus, show that the force per unit mass induced at a distance r from a straight infinite wire of unit mass per unit length is

$$F(r) = -\frac{r}{4\pi}\int_{-\infty}^{\infty} \frac{dz}{(z^2 + r^2)^{3/2}}$$

Integrating, and noting that $F(r) = -\phi'(r)$, obtain $\phi = (1/2\pi) \ln r$.

5.2. Verify the evaluation of the integral presented in (5.23). *Hint:* Complete the square in the exponent by expressing

$$i\omega(\xi - x) - \frac{\omega^2(t - \tau)}{\kappa} = -\frac{[(\omega - c)^2 - c^2](t - \tau)}{\kappa}$$

where $c = i\kappa(\xi - x)/2(t - \tau)$. Then, setting $\zeta = (\omega - c)\sqrt{(t - \tau)/\kappa}$, show that

$$U = \frac{H(t - \tau)}{2\pi\sqrt{\kappa(t - \tau)}} e^{-\kappa(\xi - x)^2/4(t - \tau)} \int_{-\infty - ia}^{\infty - ia} e^{-\zeta^2}\, d\zeta$$

where $a = (\xi - x)\sqrt{\kappa/4(t - \tau)}$. Applying Cauchy's theorem, show that the limits can be changed to $-\infty$, ∞ and, finally, recall that the integral of $\exp(-\zeta^2)$ from $-\infty$ to $+\infty$ is $\sqrt{\pi}$.

5.3. Applying Fourier and/or Laplace transforms to (5.24), deduce the result (5.25). Express (5.25) in terms of Heaviside functions and verify, a posteriori, that it does in fact satisfy (5.24).

5.4. It was pointed out that $J_0(kr)$ and $Y_0(kr)$ are linearly independent solutions of (5.13). Another set consists of the *Hankel functions of zeroth order*, $H_0^{(1)}(kr)$ and $H_0^{(2)}(kr)$, which are related to J_0 and Y_0 according to $H_0^{(1)}(z) = J_0(z) + iY_0(z)$ and $H_0^{(2)}(z) = J_0(z) - iY_0(z)$. Noting the asymptotic behavior

$$H_0^{(1)}(z) \sim 2i \ln z \quad \text{and} \quad H_0^{(2)}(z) \sim -2i \ln z \qquad \text{as } z \to 0$$

deduce additional principal solutions of the form $AH_0^{(1)}(kr)$ and $BH_0^{(2)}(kr)$, where A and B are suitably chosen constants.

5.5. Sketch the principal solutions (5.23) and (5.25) as functions of ξ, at $\tau = 0, 1, 2, 3, 4$, for $x = 1$ and $t = 2$.

5.6. Integral transform methods are of considerable utility in finding principal solutions. Let us consider, for example, the *modified Helmholtz operator*, $\nabla^2 - k^2$ ("modified" because of the minus sign). The form of ∇^2 in polar coordinates lends itself to a **Hankel transform.** [The Hankel transform of order n, and its inverse, are given by

$$\bar{f}(\rho) = \int_0^\infty rJ_n(r\rho)f(r)\,dr, \qquad f(r) = \int_0^\infty \rho J_n(r\rho)\bar{f}(\rho)\,d\rho$$

respectively, where $n \geq -1/2$.] Starting with

$$\frac{1}{r}(rU_r)_r - k^2U = \delta(\xi - x, \eta - y) \tag{A}$$

compute U by means of a Hankel transform of order zero. *Hint:* Multiply (A) by $rJ_0(r\rho)\,dr\,d\theta$ and integrate on r from 0 to ∞, and on θ from 0 to 2π. *Note carefully* that

$$\iint \delta(\xi - x, \eta - y)rJ_0(r\rho)dr\,d\theta$$

$$= \iint \delta(\xi - x, \eta - y)J_0(r\rho)\,d\xi\,d\eta = J_0(0) = 1$$

Thus obtain $\bar{U} = -1/2\pi(\rho^2 + k^2)$, and invert by means of the known formula

$$\int_0^\infty \frac{\rho J_0(r\rho)}{\rho^2 + k^2}\,d\rho = K_0(kr)$$

[I_0 and K_0 are the *modified Bessel functions of the first and second kind, and order zero.* The general solution of $xy'' + y' - k^2xy = 0$ is $y = AI_0(kx) + BK_0(kx)$.]

5.7. Obtain the solution $U = -K_0(kr)/2\pi$, to Exercise 5.6, by an entirely different method. [Note that $I_0(z) \sim 1$ and $K_0(z) \sim -\ln z$ as $z \to 0$.]

5.8. Denote the right-hand sides of (5.3) and (5.5) as w_k and U_k, respectively; as $k \to \infty$, $w_k \to \delta$ in the sense of generalized functions, and $U_k \to U$ [given by (5.6)] in the ordinary sense. Now, $\mathbf{L}^*U_k = w_k$, and $\lim \mathbf{L}^*U_k = \lim w_k = \delta$, so that U satisfies (5.1), as claimed, only if $\lim \mathbf{L}^*U_k = \mathbf{L}^*(\lim U_k)$ or, $\mathbf{L}^*U_k \to \mathbf{L}^*U$. Verify this equality. *Hint:* Show that $(\mathbf{L}^*U_k, h) \to (\mathbf{L}^*U, h)$ or $(U_k - U, \mathbf{L}h) \to 0$, and use the *Schwarz inequality* $|(f, g)| \leq \|f\|\|g\|$ where $\|f\| \equiv \sqrt{(f, f)}$.

6. GREEN'S FUNCTION METHOD FOR THE LAPLACE OPERATOR

As our first application of the Green's function method, let us consider the solution of the Poisson equation (1.5), $\nabla^2 u = \phi$, throughout a given region $\mathscr{S}$, with values $u = f$, say, prescribed on the boundary $\mathscr{C}$; see Fig. 6.1. This, of course, includes the Laplace equation, where $\phi = 0$, as a special case.

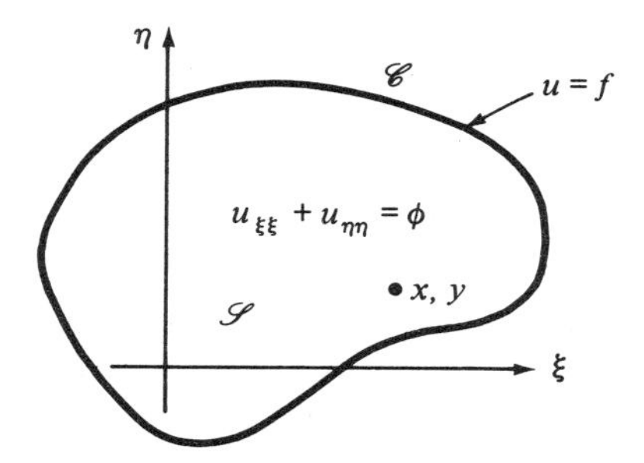

FIGURE 6.1 Summary of the boundary value problem.

As outlined in Section 4, our starting point is equation (2.7), with "v" chosen to be the Green's function G. Requiring that

$$\mathbf{L}^*G = G_{\xi\xi} + G_{\eta\eta} = \delta(\xi - x, \eta - y) \tag{6.1}$$

(2.7) reduces to

$$\iint_{\mathscr{S}} G\phi \, d\sigma = \int_{\mathscr{C}} (Gu_n - uG_n) \, ds + u(x, y) \tag{6.2}$$

Now, whereas u is prescribed on $\mathscr{C}$, the normal derivative u_n is not. To remove the unwelcome Gu_n term we therefore require that $G = 0$ on $\mathscr{C}$. This, together with (6.1), completes the specification of the boundary value problem on G. Provided, of course, that we can in fact determine G, the solution then follows from (6.2) as

$$u(x, y) = \int_{\mathscr{C}} f G_n \, ds + \iint_{\mathscr{S}} G\phi \, d\sigma \tag{6.3}$$

Notice that we use x, y to denote the fixed "field point," and ξ, η temporarily as the independent variables merely so that we end up with $u(x, y)$ in the final equation (6.3), rather than $u(\xi, \eta)$.

As discussed in Section 4, it is convenient to seek G in the form

$$G(\xi, \eta; x, y) = U(\xi, \eta; x, y) + g(\xi, \eta; x, y) = \frac{1}{2\pi} \ln r + g \tag{6.4}$$

where

$$r = \sqrt{(\xi - x^2) + (\eta - y)^2} \tag{6.4a}$$

and

$$\mathbf{L}^* g = \nabla^2 g = g_{\xi\xi} + g_{\eta\eta} = 0 \tag{6.5}$$

throughout $\mathscr{S}$. Finally, $G = 0$ on $\mathscr{C}$ implies, from (6.4), that

$$g = -\frac{1}{2\pi} \ln r \tag{6.6}$$

on $\mathscr{C}$. Equation (6.5) and the boundary condition (6.6) constitute the final boundary value problem with which we must contend. To proceed further, we must specify the region $\mathscr{S}$.

Example 1. *Circular Disk.* Consider, first, the case where $\mathscr{S}$ is a disk of radius R, with its center at the origin. A reasonable line of approach apparently is to change to polar coordinates,

$$\begin{aligned} \xi &= \tilde{\rho} \cos \tilde{\vartheta}, & x &= \rho \cos \vartheta \\ \eta &= \tilde{\rho} \sin \tilde{\vartheta}, & y &= \rho \sin \vartheta \end{aligned} \tag{6.7}$$

and use separation of variables on the resulting Laplace equation. This leads to the expression

$$g = \sum_{n=0}^{\infty} \left(\frac{\tilde{\rho}}{R}\right)^n [a_n \cos n\tilde{\vartheta} + b_n \sin n\tilde{\vartheta}] \tag{6.8}$$

where the a_n's and b_n's are computed by application of the boundary condition (6.6). Changing $r = [(\xi - x)^2 + (\eta - y)^2]^{1/2}$ in (6.6) to the polar variables defined above, and setting $\tilde{\rho} = R$, the boundary condition becomes

$$-\frac{1}{4\pi} \ln [R^2 + \rho^2 - 2R\rho \cos (\tilde{\vartheta} - \vartheta)] = \sum_{n=0}^{\infty} [a_n \cos n\tilde{\vartheta} + b_n \sin n\tilde{\vartheta}] \tag{6.9}$$

That is, we expand the left-hand side of (6.9) in a Fourier series in $\tilde{\vartheta}$, and equate its cosine and sine coefficients to a_n and b_n respectively.

Instead of completing these calculations, let us develop two simpler and more elegant lines of approach, using the method of images, and conformal mapping.

The Method of Images. Our starting point is the basic geometrical statement that the circle $\mathscr{C}$ (see Fig. 6.2), of radius R, is the locus of all points P such that

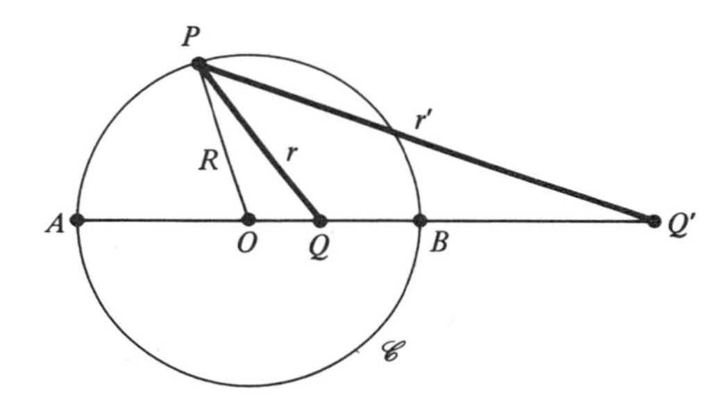

FIGURE 6.2 Image geometry.

$$QP/Q'P = \text{constant, say } \kappa \tag{6.10}$$

Given the radius R, and a distance $OQ < R$, we can use (6.10) to determine both OQ' and the constant. That is, letting P coincide with points A and B, in turn, we have

$$\frac{QP}{Q'P} = \frac{R + OQ}{R + OQ'} = \frac{R - OQ}{OQ' - R} = \kappa \tag{6.11}$$

Solving these last two equations for OQ' and κ, we have

$$\begin{aligned} OQ' &= \frac{R^2}{OQ} \\ \kappa &= \frac{OQ}{R} \end{aligned} \tag{6.12}$$

It follows that the quantity

$$\frac{1}{\kappa}\frac{QP}{Q'P} = \frac{R}{OQ}\frac{QP}{Q'P}$$

is equal to unity on the circle $\mathscr{C}$. Now let P be an arbitrary point *inside* the circle. If we take Q to be our x, y field point, and P to be the variable ξ, η point, then QP is the distance r referred to in (6.4), and $OQ = \rho$. Defining $Q'P \equiv r'$, as well, we see that

$$\frac{1}{2\pi}\ln\left(\frac{R}{OQ}\frac{QP}{Q'P}\right) = \frac{1}{2\pi}\ln r - \frac{1}{2\pi}\ln r' + \frac{1}{2\pi}\ln\frac{R}{\rho} \tag{6.13}$$

is equal to zero on $\mathscr{C}$, since the argument of the logarithm on the left side was shown to be unity on $\mathscr{C}$. Furthermore, it is *harmonic* (i.e., satisfies Laplace's equation) throughout the disk, except at Q, where it behaves like $(1/2\pi)\ln r$; to see this, note that $\ln r'$ is harmonic everywhere except at Q', which is *outside* the disk. But these are precisely the conditions required of G. Since

$$r = \sqrt{\tilde{\rho}^2 + \rho^2 - 2\tilde{\rho}\rho\cos(\tilde{\vartheta} - \vartheta)} \tag{6.14a}$$

$$r' = \sqrt{\left(\frac{R^2}{\rho}\right)^2 + \tilde{\rho}^2 - 2\tilde{\rho}\left(\frac{R^2}{\rho}\right)\cos(\tilde{\vartheta} - \vartheta)} \tag{6.14b}$$

we have

$$\begin{aligned} G &= \frac{1}{4\pi}\ln[\tilde{\rho}^2 + \rho^2 - 2\tilde{\rho}\rho\cos(\tilde{\vartheta} - \vartheta)] \\ &\quad -\frac{1}{4\pi}\ln\left[\left(\frac{R^2}{\rho}\right)^2 + \tilde{\rho}^2 - 2\tilde{\rho}\left(\frac{R^2}{\rho}\right)\cos(\tilde{\vartheta} - \vartheta)\right] + \frac{1}{2\pi}\ln\left(\frac{R}{\rho}\right) \end{aligned} \tag{6.15}$$

With $ds = R\,d\tilde{\vartheta}$ and $d\sigma = \tilde{\rho}\,d\tilde{\rho}\,d\tilde{\vartheta}$, the solution (6.3) becomes

$$u = \int_0^{2\pi} fG_n R\,d\tilde{\vartheta} + \int_0^{2\pi}\int_0^R G\phi\tilde{\rho}\,d\tilde{\rho}\,d\tilde{\vartheta} \tag{6.16}$$

where G is given by (6.15), G_n by

$$G_n = G_{\tilde{\rho}}|_{\tilde{\rho}=R} = \frac{1}{2\pi R}\frac{R^2 - \rho^2}{R^2 + \rho^2 - 2R\rho\cos(\tilde{\vartheta} - \vartheta)}$$

and f and ϕ are both prescribed. For the Laplace equation, where $\phi = 0$, (6.16) reduces to the well-known **Poisson integral formula**[12]

$$\begin{aligned} u(\rho, \vartheta) &= \frac{1}{2\pi}\int_0^{2\pi}\frac{R^2 - \rho^2}{R^2 + \rho^2 - 2R\rho\cos(\tilde{\vartheta} - \vartheta)}f(\tilde{\vartheta})\,d\tilde{\vartheta} \\ &\equiv \int_0^{2\pi} P(\tilde{\vartheta} - \vartheta, \rho)f(\tilde{\vartheta})\,d\tilde{\vartheta} \end{aligned} \tag{6.17}$$

where P is the so-called "Poisson kernel."

Physical interpretation of our Green's function, given by (6.15) or, equivalently, by the right-hand side of (6.13), is both interesting and informative. In terms of gravitational potential, we observe that the first term on the right side of (6.13) is the potential induced by a point unit mass located at the x, y field point. The second term is the potential induced by a "negative" point unit mass located at the *image point* outside the circle, as shown in Fig. 6.3. Together, these point masses set up a potential field such that our

[12]We denote $u(\rho, \vartheta)$ for brevity; strictly speaking, we mean $u[x(\rho, \vartheta), y(\rho, \vartheta)]$.

circle $\mathscr{C}$ is an equipotential curve, i.e., a curve on which the potential is constant, as sketched in Fig. 6.4. The constant potential induced on $\mathscr{C}$ is not zero however, so that a third field, simply a uniform potential, must be superimposed in order to cancel this value exactly. That explains the third and final term, $(1/2\pi) \ln (R/\rho)$, in (6.13).

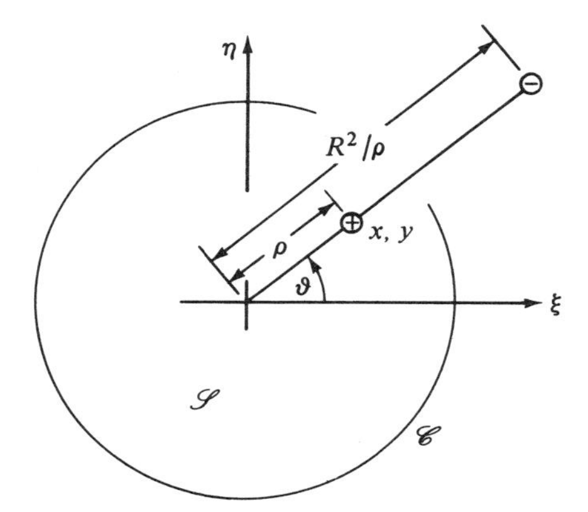

FIGURE 6.3 Point mass interpretation of *G*.

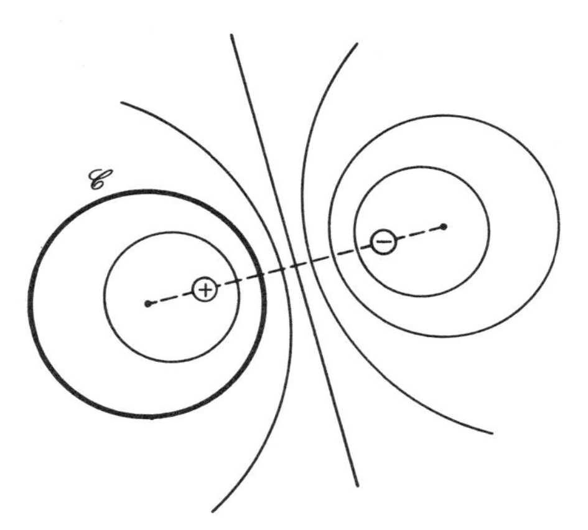

FIGURE 6.4 Equipotentials of the two point masses.

We note that the image point Q' is sometimes called the "reciprocal" or "inverse" point, since Q and Q' are at ρ, ϑ and $1/\rho$, ϑ respectively, for the case where $R = 1$.

In this example the image system was not particularly obvious. There are, as we shall see, many important cases where its construction is, however, both obvious and simple.

Conformal Mapping. Another elegant approach, which involves *conformal mapping*,[13] is available for the special case of the two-dimensional Laplace differential operator with *Dirichlet boundary conditions* (i.e., the function values prescribed on the boundary $\mathscr{C}$). We assert that G is then given by

$$G = \frac{1}{2\pi}\ln|w(z)| = \frac{1}{2\pi}\mathscr{R}\,[\ln w(z)] \tag{6.18}$$

where $w(z)$ is an analytic function of the complex variable $z = \xi + i\eta$ throughout the region $\mathscr{S}$ plus its boundary $\mathscr{C}$, which maps $\mathscr{S}$ conformally and one-one onto the unit disk, such that $x + iy \equiv \alpha$ goes into the origin and $\mathscr{R}$ denotes "the real part of." The proof is as follows: Since $w(z)$ is analytic, and maps α into the origin, it must be of the form

$$w = (z - \alpha)^n P(z)$$

where n is a positive integer and $P(z)$ is analytic and nonzero over $\mathscr{S} + \mathscr{C}$. [If P were zero at some point $z \neq \alpha$ then this point would also map into the origin, and the mapping would not be one-one. If it were zero at α, then it would contain a $(z - \alpha)^m$ factor, which could be absorbed into the $(z - \alpha)^n$ factor.] Conformality demands, further, that $n = 1$, since $n \geq 2$ would imply that $w'(\alpha) = 0$. Thus $w = (z - \alpha)P(z)$, and

$$\begin{aligned} G &= \frac{1}{2\pi}\mathscr{R}\{\ln[(z - \alpha)P(z)]\} \\ &= \frac{1}{2\pi}\mathscr{R}[\ln(z - \alpha)] + \frac{1}{2\pi}\mathscr{R}[\ln P(z)] \end{aligned} \tag{6.19}$$

Now, if a function $F(z)$ is analytic, then $\ln F(z)$ is analytic wherever $F(z) \neq 0$. It follows that the second term on the right-hand side of (6.19) is the real part of an analytic function, and is hence harmonic over $\mathscr{S} + \mathscr{C}$. Similarly, the first term is harmonic except at α. Thus, G is harmonic over $\mathscr{S} + \mathscr{C}$, except at α, where

$$G \sim \frac{1}{2\pi}\ln|z - \alpha| \tag{6.20}$$

as required, since $|z - \alpha| = r$; c.f. equation (6.4). Finally, we note that $|w(z)| = 1$ on $\mathscr{C}$, since $\mathscr{C}$ maps into the unit circle, so that $G = 0$ on $\mathscr{C}$, from (6.18), as required.

An interesting consequence of this result is that we are assured of the *existence* of a Green's function for all domains $\mathscr{S}$ which can be mapped con-

[13]Our discussion will assume of the reader a basic knowledge of complex variable theory, approximately at the level of R. V. Churchill's *Complex Variables and Applications*, 2nd ed., McGraw-Hill Book Company, New York, 1960. It can, however, be skipped without loss of continuity if the reader is not sufficiently familiar with this material.

formally onto the unit disk. That these include all simply connected domains with piecewise smooth boundaries is Riemann's famous mapping theorem.[14] Unfortunately, assurance that a suitable mapping exists is not of much help in *finding* it.[15]

To apply this method to the present example, we note that the bilinear transformation

$$w(z) = \frac{R(z - \alpha)}{R^2 - \bar{\alpha}z} \tag{6.21}$$

where the bar on α denotes complex conjugate, maps the disk $\mathscr{S}$ of radius R into the unit disk, both conformally and one-one, with α going into the origin. Thus, with $z = \tilde{\rho}\exp(i\tilde{\vartheta})$ and $\alpha = \rho\exp(i\vartheta)$, we have

$$G = \frac{1}{2\pi}\ln\left|\frac{R\tilde{\rho}e^{i\tilde{\vartheta}} - R\rho e^{i\vartheta}}{R^2 - \rho\tilde{\rho}e^{i(\tilde{\vartheta}-\vartheta)}}\right| \tag{6.22}$$

which can be shown (Exercise 6.1) to agree with our previous result (6.15).

COMMENT 1. Strictly speaking, the validity of our basic equation (2.7) is subject to question when we set "v" $= G$, because of G's singular behavior within the domain of integration. Following more rigorous lines, however, we can in fact show that our result (6.2) is valid. Specifically, we apply (2.7) not to the region $\mathscr{S}$, but rather to $\mathscr{S}'$, which is identical to $\mathscr{S}$, but with a disk of arbitrarily small radius ϵ deleted at the x, y location; see Fig. 6.5. With the singular point removed, (2.7) *is* rigorously correct, and becomes

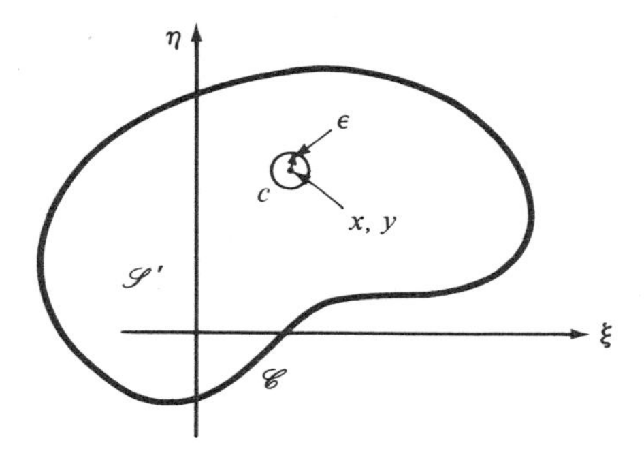

FIGURE 6.5 Rigorous derivation of Eq. (6.2).

[14] C. Carathéodory, *Conformal Representation*, Cambridge University Press, New York, 1932.

[15] Dictionaries of conformal mappings are available and are of considerable help; e.g., H. Kober, *Dictionary of Conformal Representations*, Dover Publications, New York, 1952, and Z. Nehari, *Conformal Mapping*, McGraw-Hill Book Company, New York, 1952.

$$\iint_{\mathscr{S}'} G\nabla^2 u \, d\sigma = \int_{\mathscr{C}} (Gu_n - uG_n)\, ds + \int_c (Gu_n - uG_n)\, ds + \iint_{\mathscr{S}'} u\nabla^2 G \, d\sigma \tag{6.23}$$

since the boundary of $\mathscr{S}'$ is actually $\mathscr{C} + c$. Consider the integral over c: $G = O(\ln \epsilon)$, $u_n = O(1)$,[16] and the length of the circuit is $2\pi\epsilon$, so that

$$\int_c Gu_n \, ds = O(\epsilon \ln \epsilon) \tag{6.24}$$

which tends to zero as $\epsilon \to 0$. In addition, $u \sim u(x, y)$ and

$$G_n = -G_r \sim -\frac{1}{2\pi r}\bigg|_{r=\epsilon} = -\frac{1}{2\pi\epsilon} \tag{6.25}$$

so that

$$\int uG_n \, ds \sim -\frac{u(x, y)}{2\pi\epsilon} \int_c ds = -u(x, y) \tag{6.26}$$

Finally, the last term is zero in (6.23) since $\nabla^2 G = 0$ in $\mathscr{S}'$, and the left-hand side becomes

$$\iint_{\mathscr{S}'} G\nabla^2 u \, d\sigma = \iint_{\mathscr{S}} G\phi \, d\sigma \tag{6.27}$$

since $\nabla^2 u = \phi$ and the integral of $G\phi$ over the disk $\mathscr{S} - \mathscr{S}'$ is of order $O(\epsilon^2 \ln \epsilon)$; $O(\epsilon^2)$ from the disk area, and $O(\ln \epsilon)$ from G. With these results, (6.23) does indeed reduce to our equation (6.2), as claimed.

Rather than worry about the rigor of each and every step in the solution, it is often best to proceed formally all the way. If it can then be verified that the "solution" does in fact satisfy all the required conditions, then the question of rigor of the various intermediate steps is of no concern. (In fact, the solution often turns out to be valid under somewhat milder conditions, on the various prescribed quantities, than would have accrued if we had demanded rigor at each step of the way.)

To illustrate the procedure, let us try to verify that the Poisson integral formula does in fact satisfy the required conditions, $\nabla^2 u = 0$ in the unit disk and $u = f$ on its edge. First, let us sharpen our problem statement a bit. Actually, we require that $\nabla^2 u = 0$ in the *interior* of the disk, $r < R$, and not on its boundary, $r = R$. [This is important in boundary and initial value problems so that nonanalytic (for example, discontinuous) boundary and initial conditions can be accommodated.] By $u = f$ on $\rho = R$, then, we shall mean $\lim u(\rho, \vartheta) = f(\vartheta)$ as $\rho \to R$. That this boundary condition is satisfied follows

[16] That is, u_n is bounded; recall the definition of the order-of-magnitude notation in Footnote 14 of PART I.

immediately from the delta function behavior of the Poisson kernel, which was noted in Exercise 3.3 of PART I and left for the reader to prove. That $\nabla^2 u$ does equal 0 in $\rho < R$ follows from the fact that $\nabla^2 P = P_{\rho\rho} + \rho^{-1} P_\rho + \rho^{-2} P_{\vartheta\vartheta} = 0$ in $\rho < R$, differentiation under the integral sign being permissible because the required partials of P are all continuous functions of the parameters ρ, ϑ and the integration variable $\tilde{\vartheta}$ over $\rho < R$, $0 \leq \vartheta \leq 2\pi$, $0 \leq \tilde{\vartheta} \leq 2\pi$.

COMMENT 2. Interchanging ρ with $\tilde{\rho}$ and ϑ with $\tilde{\vartheta}$ in the Green's function (6.14), we find (Exercise 6.2) that $G(\tilde{\rho}, \tilde{\vartheta}; \rho, \vartheta) = G(\rho, \vartheta; \tilde{\rho}, \tilde{\vartheta})$ so that G is *symmetric*. Let us examine the source of this reciprocity. Our argument will be analogous to the corresponding one for ordinary differential operators, in Section 4 of PART I (Example 1, Comment 3).

Recall that G satisfies a system of the form

$$\mathbf{L}^* G(\xi, \eta; x, y) = \delta(\xi - x, \eta - y), \text{ plus suitable homogeneous boundary conditions, say B. C.} \tag{6.28}$$

Denoting the operator of (6.28) (i.e., $\mathbf{L}^*$ plus B.C.) as $\mathscr{G}$, we note that associated with $\mathscr{G}$ is its adjoint $\mathscr{G}^*$, consisting of $(\mathbf{L}^*)^* = \mathbf{L}$ plus certain other homogeneous boundary conditions, say B. C.*. Thus, corresponding to $\mathscr{G}^*$ there is a so-called adjoint Green's function, say G^*, which satisfies

$$(\mathbf{L}^*)^* G^*(\xi, \eta; x_0, y_0) = \mathbf{L} G^*(\xi, \eta; x_0, y_0) = \delta(\xi - x_0, \eta - y_0) \text{ plus suitable homogeneous boundary conditions, B. C.*} \tag{6.29}$$

where we have introduced the zero subscripts on x, y simply because x_0, y_0 in (6.29) need not be the same as x, y in (6.28).

Now, taking u and v in (2.12) to be $G^*(\xi, \eta; x_0, y_0)$ and $G(\xi, \eta; x, y)$ respectively, and using (2.13), (6.28), and (6.29), we have

$$\iint_{\mathscr{S}} \delta(\xi - x_0, \eta - y_0) G(\xi, \eta; x, y)\, d\sigma$$
$$= \iint_{\mathscr{S}} G^*(\xi, \eta; x_0, y_0) \delta(\xi - x, \eta - y)\, d\sigma$$

or

$$G(x_0, y_0; x, y) = G^*(x, y; x_0, y_0) \tag{6.30}$$

In the present example $\mathscr{G}$ consists of $\mathbf{L}^* = \nabla^2$ plus the boundary condition $G = 0$ on $\mathscr{C}$. Now, with u and v taken to be G^* and G, respectively, in (2.7), we see that G^* must be zero on $\mathscr{C}$ in order for the boundary term to drop out, as implied by (2.12). Thus $\mathscr{G}^*$ consists of $\mathbf{L} = \nabla^2$ plus the boundary condition $G^* = 0$ on $\mathscr{C}$, so that $\mathscr{G}^* = \mathscr{G}$, and (6.30) reduces to the simple symmetry condition

$$G(x_0, y_0; x, y) = G(x, y; x_0, y_0) \tag{6.31}$$

COMMENT 3. We have seen that for the case where $\mathbf{L} = \nabla^2$, U may be regarded as the potential induced by a point unit mass located at x, y within $\mathscr{S}$, and that g is an additional potential which (if we are lucky) is derivable from a simple image system. The image system consists of a distribution of mass, either discrete or continuous (or partly discrete and partly continuous), located *outside* the region $\mathscr{S}$. We will now demonstrate that g (in fact, any function harmonic in $\mathscr{S}$) may be regarded, alternatively, as the potential induced by a suitable distribution of singularities over the *boundary* $\mathscr{C}$.[17] First, we define the notion of a mass **dipole**, or **doublet**: Consider (Fig. 6.6) the potential $u(x', y')$ induced at x', y' by a pair of point masses, one of strength $+\kappa$ at $\epsilon, 0$ and the other of strength $-\kappa$ at 0, 0. With

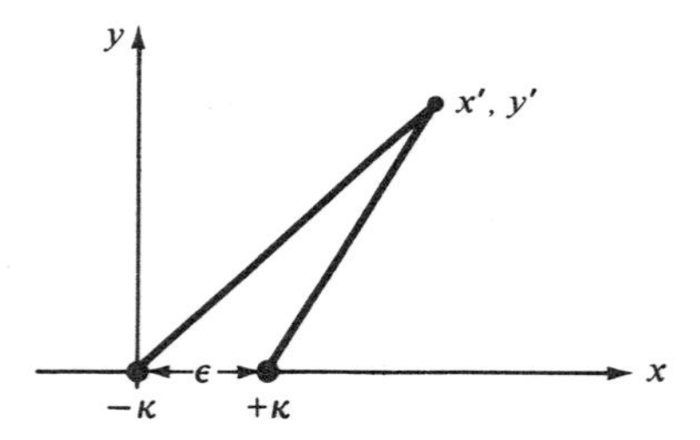

FIGURE 6.6 Definition of doublet.

$$U(x', y'; x, y) = \frac{1}{2\pi} \ln [x' - x)^2 + (y' - y)^2]^{1/2} \tag{6.32}$$

the potential at x', y' due to a positive unit mass at x, y it follows that

$$\begin{aligned} u(x', y') &= \kappa U(x', y'; \epsilon, 0) - \kappa U(x', y'; 0, 0) \\ &= \kappa\epsilon \left\{ \frac{U(x', y'; \epsilon, 0) - U(x', y'; 0, 0)}{\epsilon} \right\} \end{aligned} \tag{6.33}$$

Letting $\epsilon \to 0$ and $\kappa \to \infty$ such that the product $\kappa\epsilon = \text{constant} = \nu$, say, (6.33) becomes

$$u(x', y') = \nu \frac{\partial U}{\partial x} \tag{6.34}$$

[17] Although this method will not be pursued further here, we feel it is worth mentioning since it is of important practical interest in potential theory. It has, for example, been exploited extensively in the fields of aero- and hydrodynamics.

The result is said to be a "dipole," or "doublet," of strength ν, oriented in the positive x direction and located at the origin. It is important to note that the derivative $\partial/\partial x$ always coincides with the orientation of the dipole.

Now, setting $u = U(\xi, \eta; x, y)$ and $v = g(\xi, \eta)$ in *Green's theorem,*

$$\iint_{\mathscr{S}} (u\nabla^2 v - v\nabla^2 u)\, d\sigma = \int_{\mathscr{C}} \left\{u\frac{\partial v}{\partial n} - v\frac{\partial u}{\partial n}\right\} ds \tag{6.35}$$

where $\nabla^2 g = 0$ in $\mathscr{S}$ and $\nabla^2 U = \delta(\xi - x, \eta - y)$, we have

$$0 - g(x, y) = \int_{\mathscr{C}} \left\{U(\xi, \eta; x, y)\frac{\partial g}{\partial n} - g\frac{\partial}{\partial n}U(\xi, \eta; x, y)\right\} ds \tag{6.36}$$

or, since $U(\xi, \eta; x, y) = U(x, y; \xi, \eta)$,

$$g(x, y) = \int_{\mathscr{C}} \left\{-\frac{\partial g}{\partial n}U(x, y; \xi, \eta) + g\frac{\partial}{\partial n}U(x, y; \xi, \eta)\right\} ds \tag{6.37}$$

To interpret the two terms in the integrand, note that $U(x, y; \xi, \eta)$ is the potential induced at x, y by a point unit mass at ξ, η on $\mathscr{C}$. Similarly, $\partial U(x, y; \xi, \eta)/\partial n$ is the potential induced at x, y by a point unit doublet at ξ, η on $\mathscr{C}$, oriented normal to $\mathscr{C}$.

Thus, $g(x, y)$ can in fact be interpreted as the gravitational potential induced at the interior point x, y by a distribution of masses and dipoles, of strength $-\partial g/\partial n$ and g respectively, over $\mathscr{C}$. The nonuniqueness of this singularity representation is established in Exercise 6.3.

Example 2. *Half-Plane.* Now consider the case where $\mathscr{S}$ consists of the upper half-plane, $\eta > 0$. This time, construction of the image system is trivial; it consists simply of a negative point unit mass at $x, -y$ as shown in Fig. 6.7. Surely G will then be zero on $\eta = 0$ by antisymmetry. Thus,

$$G = \frac{1}{4\pi}\ln\,[(\xi - x)^2 + (\eta - y)^2] - \frac{1}{4\pi}\ln\,[(\xi - x)^2 + (\eta + y)^2] \tag{6.38}$$

Clearly this does satisfy all the required conditions. With $G_n = -G_\eta|_{\eta=0}$, the solution is provided by (6.3),

$$u(x, y) = \frac{y}{\pi}\int_{-\infty}^{\infty} \frac{f(\xi)}{(\xi - x)^2 + y^2}\, d\xi$$
$$+ \frac{1}{4\pi}\int_0^{\infty}\int_{-\infty}^{\infty} \ln\left[\frac{(\xi - x)^2 + (\eta - y)^2}{(\xi - x)^2 + (\eta + y)^2}\right]\phi(\xi, \eta)\, d\xi\, d\eta \tag{6.39}$$

The Green's function (6.38) could also have been found easily using conformal mapping, with $w(z) = (z - \alpha)/(z - \bar{\alpha})$.

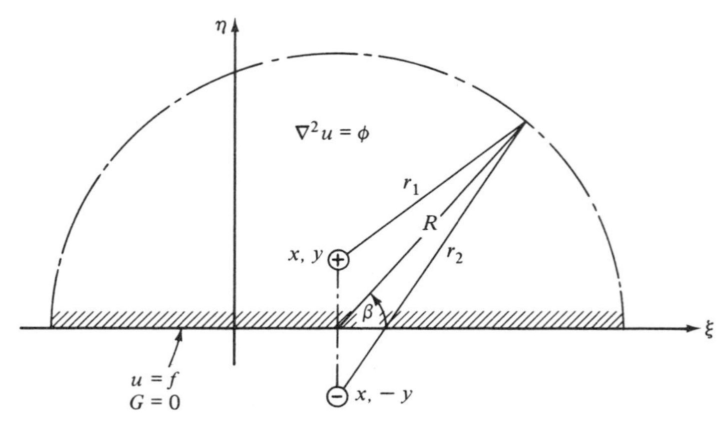

FIGURE 6.7 Image system for half-plane.

COMMENT 1. Actually, our region $\mathscr{S}$ should be closed, as sketched in Fig. 6.1. If we close it by means of a large semicircle centered at $x, 0$ for convenience (Fig. 6.7), and consider the limit as $R \to \infty$, then we see from (6.2) that our result (6.39) tacitly assumes that u satisfies additional boundary conditions "at infinity" such that the integral

$$\int (Gu_n - uG_n)\, ds \tag{6.40}$$

over the semicircle tends to zero as $R \to \infty$. On the semicircle,

$$\begin{aligned} G &= \frac{1}{2\pi} \ln r_1 - \frac{1}{2\pi} \ln r_2 \\ &\sim \frac{1}{2\pi} \ln (R - y \sin \beta) - \frac{1}{2\pi} \ln (R + y \sin \beta) \\ &\sim -\frac{y \sin \beta}{\pi R} = O\left(\frac{1}{R}\right) \end{aligned} \tag{6.41}$$

since $\ln (z + a) \sim \ln z + a/z$ as $z \to \infty$. It follows that $G_n = G_R = O(1/R^2)$ on the semicircle, so that (6.40) must be of order

$$O\left(\frac{1}{R} \cdot u_R \cdot \pi R\right) + O\left(u \cdot \frac{1}{R^2} \cdot \pi R\right)$$

This will in fact tend to zero if both $u_R \to 0$ and $u/R \to 0$ as $R \to \infty$; these are equivalent to the single requirement $u = O(R^\alpha)$, where $\alpha < 1$.

Stipulation of appropriate boundary conditions at infinity is an important part of problems in infinite domains. Quoting Friedrichs,[18] "*conditions are ap-*

[18] K. O. Friedrichs, *Methods of Mathematical Physics* (Lectures given during 1947–1948), New York University Institute of Mathematical Sciences, New York, 1955.

propriate if they are strong enough so that at most one solution can satisfy them, and weak enough so that there exists at least one solution satisfying them." As a simple illustration of this point he notes that the solution of the two-dimensional Laplace equation in the region $r > 1$ subject to the condition $u = 100$ on $r = 1$ is $u = 100 + C \ln r$, where r is the radial distance from the origin. Suppose we require one of the following as our condition at infinity: (a) $u \to 0$, (b) $u = O(1)$, (c) $u = O(r)$. Condition (a) cannot be satisfied for any choice of C, so that *no* solution exists; (b) is satisfied if and only if $C = 0$, so that there does exist a unique solution; (c) is satisfied for *any* value of C, since $(\ln r)/r$ is bounded as $r \to \infty$, so that the solution exists but is nonunique. For a detailed discussion of this and other points related to **infinite domains**, we refer the interested reader to Friedrichs.

COMMENT 2. Observe from (6.39) that the lim $u(x, y)$ as $y \to 0$ is in fact $f(x)$, as required, since the kernel of the first integral behaves as a delta function at $\xi = x$ as $y \to 0$ [recall equation (3.3) of PART I, with $k = 1/y$], and the kernel of the second integral tends to zero.

The foregoing examples have all been formulated with *Dirichlet boundary conditions*, that is $u = f$ prescribed on $\mathscr{C}$ (recall Fig. 6.1). Examples with *Neumann boundary conditions*, that is where the normal derivative u_n is prescribed on $\mathscr{C}$, are given in Exercises 6.9 and 6.10. Let us consider now the more general *mixed* boundary condition

$$\alpha u + \beta u_n = f \text{ on } \mathscr{C} \tag{6.42}$$

with α, β, and f prescribed; for the special cases $\alpha = 1$, $\beta = 0$ and $\alpha = 0$, $\beta = 1$ reduction to Dirichlet and Neumann type conditions results.

Let us return to (6.2), more specifically the boundary integral, in order to determine what boundary conditions need to be imposed on G for this case. If $\beta \neq 0$ over $\mathscr{C}$, then with the help of (6.42) we can express

$$Gu_n - uG_n = G\left(\frac{f - \alpha u}{\beta}\right) - uG_n = G\left(\frac{f}{\beta}\right) - \left(\frac{\alpha G + \beta G_n}{\beta}\right)u \tag{6.43}$$

Now, whereas the combination $\alpha u + \beta u_n = f$ is prescribed, u (all by itself) is not. To remove the unknown, and hence unwelcome, u term, we require that G satisfy the "homogenized" mixed boundary condition

$$\alpha G + \beta G_n = 0 \text{ on } \mathscr{C} \tag{6.44}$$

With G determined, the solution then follows from (6.2), (6.43), and (6.44) as

$$u(x, y) = -\int_{\mathscr{C}} \frac{Gf}{\beta}\, ds + \iint_{\mathscr{S}} G\phi \, d\sigma \tag{6.45}$$

If β is *not* nonzero over $\mathscr{C}$ but α is, then we can express

$$Gu_n - uG_n = Gu_n - \left(\frac{f - \beta u_n}{\alpha}\right)G_n = \left(\frac{\alpha G + \beta G_n}{\alpha}\right)u_n - \left(\frac{f}{\alpha}\right)G_n \quad (6.46)$$

and again (6.44) applies since u_n is not prescribed on $\mathscr{C}$. More generally, (6.44) applies if, at all points on $\mathscr{C}$, α and β are not both zero. The solution is then given by (6.45), with $-Gf/\beta$ replaced by $G_n f/\alpha$ for that portion of $\mathscr{C}$ over which β is zero but α is not. Let us illustrate this with an example.

Example 3. *Mixed Boundary Conditions.* To emphasize the effect of the mixed boundary condition let us reconsider the same problem as in Example 2, but this time with the boundary condition

$$u_n + \alpha u = f(\xi) \text{ on } \eta = 0 \quad (6.47)$$

with α a constant, for simplicity. The fact that G must now satisfy the condition

$$G_n + \alpha G = 0 \text{ on } \eta = 0 \quad (6.48)$$

causes a surprising amount of complication. The boundary value problem on G does not admit a simple image type solution, and conformal mapping is apparently of no help because we do not have $G = 0$ on $\mathscr{C}$. Instead, we will adapt an elegant method which was developed by J. P. Moran[19] in connection with the flow field induced by a hydrodynamic source moving toward a free surface.

Noting that $G_n = -G_\eta$ on $\eta = 0$, our boundary value problem on G is as follows:

$$\mathbf{L}^*G = \nabla^2 G = \delta(\xi - x, \eta - y) \qquad \text{in } \eta > 0 \quad (6.49)$$

$$-G_\eta + \alpha G = 0 \qquad \text{on } \eta = 0 \quad (6.50)$$

Proceeding formally, we seek G in the form of an expansion in powers of α,

$$G = G_0 + \alpha G_1 + \alpha^2 G_2 + \cdots \quad (6.51)$$

Inserting this into the differential equation (6.49), and equating the coefficients of like powers of α on both sides of the equation, we have

$$\nabla^2 G_0 = \delta(\xi - x, \eta - y) \quad (6.52)$$

$$\nabla^2 G_j = 0, \qquad j \geq 1 \quad (6.53)$$

Doing the same for the boundary condition (6.50),

[19] J. P. Moran, "Image Solution for Vertical Motion of a Point Source Towards a Free Surface," *J. Fluid Mechanics*, Vol. 18, Part 2, 1964, pp. 315–320.

$$-(G_{0_\eta} + \alpha G_{1_\eta} + \cdots) + \alpha(G_0 + \alpha G_1 + \cdots) = 0 \tag{6.54}$$

so that, equating powers of α,

$$G_{0_\eta} = 0 \tag{6.55}$$

$$\left.\begin{array}{r} -G_{1_\eta} + G_0 = 0 \\ -G_{2_\eta} + G_1 = 0 \\ \vdots \end{array}\right\} \tag{6.56}$$

on $\eta = 0$. From (6.52) and (6.55) it is clear (Exercise 6.10) that G_0 can be constructed by a simple image system:

$$G_0 = U(\xi, \eta; x, y) + U(\xi, \eta; x, -y) \tag{6.57}$$

where

$$U(\xi, \eta; x, y) = \frac{1}{2\pi} \ln \sqrt{(\xi - x)^2 + (\eta - y)^2}$$

Observing that on $\eta = 0$ we have $G_0 = 2V$, where we define $V(\xi, \eta; x, y) \equiv U(\xi, \eta; x, -y)$, we can rewrite (6.56) as

$$\left.\begin{array}{r} -G_{1_\eta} + 2V = 0 \\ -G_{2_\eta} + G_1 = 0 \\ \vdots \end{array}\right\} \tag{6.58}$$

on $\eta = 0$.

Now, $\nabla^2(-G_{1_\eta} + 2V) = 0$ in $\eta > 0$ since both G_1 and V are harmonic there and, in addition, $-G_{1_\eta} + 2V = 0$ on $\eta = 0$ according to (6.58). Both conditions are satisfied if $-G_{1_\eta} + 2V$ is *identically* zero throughout the half-plane. The same is true for the combinations $-G_{2_\eta} + G_1$, $-G_{3_\eta} + G_2$, and so on. Thus, the equations (6.58) are valid not only on $\eta = 0$, but for $\eta > 0$ as well; we have "continued" the boundary conditions into the region! Integrating them in turn yields

$$\left.\begin{array}{l} G_1 = 2\int V\, d\eta \\ G_2 = \int G_1\, d\eta = 2\iint V\, d\eta\, d\eta \\ G_3 = \int G_2\, d\eta = 2\iiint V\, d\eta\, d\eta\, d\eta \\ \vdots \end{array}\right\} \tag{6.59}$$

so that

$$G = G_0 + 2\alpha\int V\,d\eta + 2\alpha^2\iint V\,d\eta\,d\eta + \cdots \tag{6.60}$$

where G_0 and V are already known. This can be put into closed form by differentiating with respect to η and noting that

$$\begin{aligned} G_\eta &= G_{0\eta} + 2\alpha V + 2\alpha^2\int V\,d\eta + \cdots \\ &= G_{0\eta} + 2\alpha V + \alpha(G - G_0) \end{aligned} \tag{6.61}$$

Thus we have the simple first order differential equation

$$G_\eta - \alpha G = G_{0\eta} + 2\alpha V - \alpha G_0 \tag{6.62}$$

Since the terms on the right-hand side are already known, the solution is simply

$$G = e^{\alpha\eta}\int_a^\eta (G_{0\eta'} + 2\alpha V - \alpha G_0)e^{-\alpha\eta'}\,d\eta' \tag{6.63}$$

where the arguments of $G_{0\eta'}$, V, and G_0 are $(\xi, \eta';\, x, y)$. Choosing the integration constant $a = \infty$, for G to be well behaved as $\eta \to \infty$, further simplification by means of integration by parts is possible,

$$\begin{aligned} G &= e^{\alpha\eta}\left\{(G_0 e^{-\alpha\eta'})\Big|_\infty^\eta + \int_\infty^\eta (\alpha G_0 + 2\alpha V - \alpha G_0)e^{-\alpha\eta'}\,d\eta'\right\} \\ &= G_0 + \left(\frac{\alpha}{\pi}\right)e^{\alpha\eta}\int_\infty^\eta \ln\sqrt{(\xi - x)^2 + (\eta' + y)^2}\,e^{-\alpha\eta'}\,d\eta' \end{aligned} \tag{6.64}$$

Finally, in order to facilitate the physical interpretation of this result, let us change the dummy variable from η' to η'' according to $\eta' = \eta - y - \eta''$. Carrying out this simple change, we arrive at

$$G = U(\xi, \eta;\, x, y) + U(\xi, \eta;\, x, -y) + \int_{-\infty}^{-y} m(\eta'')U(\xi, \eta;\, x, \eta'')\,d\eta'' \tag{6.65}$$

where

$$m(\eta'') = -2\alpha e^{\alpha(y+\eta'')} \tag{6.66}$$

The first two terms in (6.65) clearly represent positive point unit masses at x, y and $x, -y$ respectively; the integral term represents an additional *distribution* of mass, of strength $m(\eta'')$ per unit length, from $x, -y$ down to $x, -\infty$, as sketched in Fig. 6.8. Thus G *is*, in fact, expressible in terms of an image system, but the system is neither simple nor obvious! (See Exercise 6.11.)

Example 4. *Quarter-Plane.* As the final example of this section, we consider another mixed boundary value problem, namely, the Poisson equation $\nabla^2 u = \phi$ in the quarter-plane, with $u = f(\xi)$ on $\eta = 0$, and $u_n = h(\eta)$ on $\xi =$

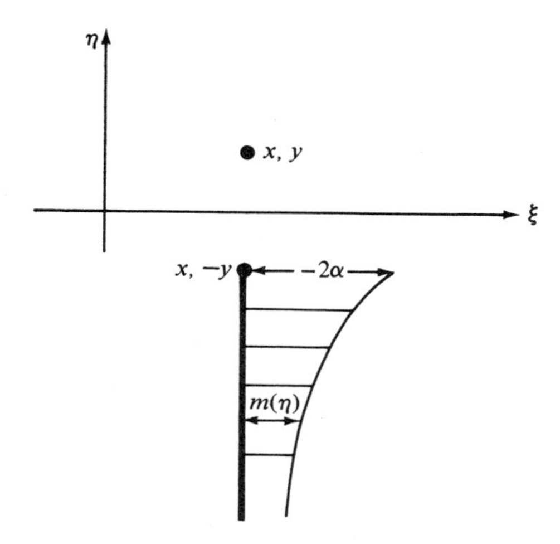

FIGURE 6.8 Image system for Example 3.

0, as shown in Fig. 6.9. The appropriate boundary value problem on G is also indicated there. By inspection, G may be constructed by means of three suitably placed unit masses, as shown: a positive mass at $-x, y$ and negative masses at $-x, -y$ and $x, -y$.

The solution is given by

$$u(x, y) = \int_{\mathscr{C}} uG_n \, ds - \int_{\mathscr{C}} Gu_n \, ds + \iint_{\mathscr{S}} G\phi \, d\sigma$$

$$= \int_0^\infty fG_n \, d\xi - \int_0^\infty Gh \, d\eta + \int_0^\infty \int_0^\infty G\phi \, d\xi \, d\eta$$

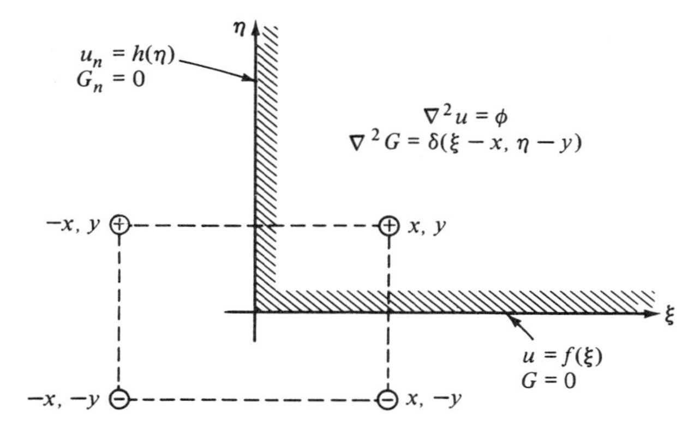

FIGURE 6.9 Image system for Example 4.

$$= -\int_0^\infty f(\xi)G_\eta(\xi, 0; x, y)\, d\xi - \int_0^\infty G(0, \eta; x, y)h(\eta)\, d\eta$$

$$+\int_0^\infty \int_0^\infty G(\xi, \eta; x, y)\phi(\xi, \eta)\, d\xi\, d\eta \tag{6.67}$$

where

$$G(\xi, \eta; x, y) = \frac{1}{4\pi} \ln \frac{[(\xi - x)^2 + (\eta - y)^2][(\xi + x)^2 + (\eta - y)^2]}{[(\xi - x)^2 + (\eta + y)^2][(\xi + x)^2 + (\eta + y)^2]} \tag{6.68}$$

It appears that not *all* mixed boundary value problems are difficult; just *most* of them.

COMMENT 1. The reader may be upset at the signs involved in the expression of the boundary integrals in (6.67), and puzzled at the "orientation" of the contour $\mathscr{C}$, since ξ runs "clockwise" from 0 to ∞, whereas η runs "counterclockwise" from 0 to ∞. As we noted earlier, the reason is that the ds increments must always be *positive*; the directions of integration must therefore be chosen accordingly.

EXERCISES

6.1. Verify that the form (6.22) does agree with (6.15).

6.2. Verify that the Green's function given by (6.15) is symmetric; that is, $G(\tilde{\rho}, \tilde{\vartheta}; \rho, \vartheta) = G(\rho, \vartheta; \tilde{\rho}, \tilde{\vartheta})$.

6.3. Establish the nonuniqueness of the singularity representation discussed in Comment 3 of Example 1. *Hint:* In deriving (6.37) we set $u = U(\xi, \eta; x, y)$ and $v = g(\xi, \eta)$ in (6.35), where g was harmonic in $\mathscr{S}$, and the point x, y was contained in $\mathscr{S}$. Alternatively, apply (6.35) to the complementary region $\mathscr{S}'$ bounded by $\mathscr{C}$; if $\mathscr{S}$ is the interior of $\mathscr{C}$ then $\mathscr{S}'$ is the exterior, and conversely. Again choose $u = U(\xi, \eta; x, y)$, where x, y is the same point (in $\mathscr{S}$) as before, but let $v = h(\xi, \eta)$ where h is harmonic in $\mathscr{S}'$. Noting that $\partial/\partial n' = -\partial/\partial n$ and that x, y is outside $\mathscr{S}'$ we obtain, from (6.35),

$$0 = \int_{\mathscr{C}} \left\{\frac{\partial h}{\partial n} U(x, y; \xi, \eta) - h\frac{\partial}{\partial n} U(x, y; \xi, \eta)\right\} ds$$

Adding this to (6.37),

$$g(x, y) = \int_{\mathscr{C}} \left\{-\frac{\partial(g - h)}{\partial n} U + (g - h)\frac{\partial U}{\partial n}\right\} ds$$

That the mass and dipole strengths over $\mathscr{C}$, $-\partial(g - h)/\partial n$ and $(g - h)$ respectively, are not unique follows from the fact that h is not uniquely specified.

6.4. Determine the Green's function for the case where $\nabla^2 u = 0$ in the region *outside* the circle of radius R, with $u(\tilde{\rho}, \tilde{\vartheta}) = f(\tilde{\vartheta})$ on the boundary $\tilde{\rho} = R$, and hence obtain the solution

$$u(\rho, \vartheta) = \frac{1}{2\pi}\int_0^{2\pi} \frac{\rho^2 - R^2}{R^2 - \rho^2 - 2R\rho\cos(\tilde{\vartheta} - \vartheta)} f(\tilde{\vartheta})\,d\tilde{\vartheta}$$

(Compare with the Poisson integral formula (6.17) for the *interior* problem.) *Hint:* First, map the $z = \xi + i\eta$ plane into a $w_1(z)$ plane by means of the inversion $w_1 = 1/z$. This sends our region into the *interior* of a disk of radius $1/R$, and $\alpha = x + iy$ into $1/\alpha$, as shown below. Finally, we must map w_1 conformally into w_2 such that the disk of radius $1/R$ maps into the *unit* disk, with $1/\alpha$ going into the origin. Recalling the form of (6.21), we see that the desired transformation is

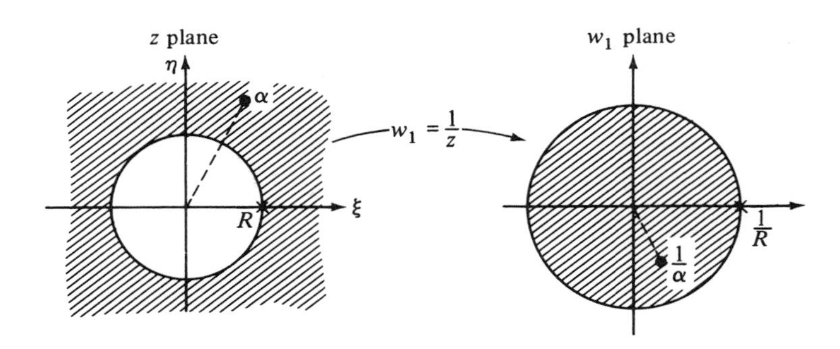

$$w_2 = \frac{(1/R)(w_1 - 1/\alpha)}{(1/R)^2 - (1/\bar{\alpha})w_1}$$

so that

$$G = \frac{1}{2\pi}\ln\left|\frac{(1/R)(1/z - 1/\alpha)}{(1/R)^2 - (1/\bar{\alpha})(1/z)}\right|$$

6.5. Repeat Exercise 6.4 using the method of images.

6.6. Show that the Green's function for the infinite wedge problem, shown below, may be constructed by placing suitable point masses at the locations $a, b, \ldots, g$.

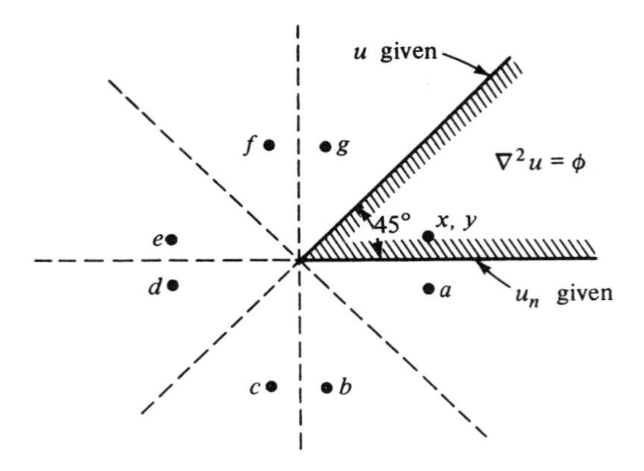

6.7. Determine the Green's function for the Poisson equation with Dirichlet boundary conditions, for a semicircular domain of radius R, using three suitably placed point masses.

6.8. What boundary conditions at infinity must be satisfied by u if (6.67) and (6.68) are to constitute a valid solution of the quarter-plane problem? (Recall Comment 1 of Example 2.)

6.9. Generalized Green's Function. Consider the Poisson equation $\nabla^2 u = \phi$ over a bounded region $\mathscr{S}$, subject to *Neumann boundary conditions* $u_n = 0$ on the boundary $\mathscr{C}$. (a) Integrating the Poisson equation over the disk, and using the divergence theorem,

$$\iint_{\mathscr{S}} \nabla^2 u \, d\sigma = \int_{\mathscr{C}} u_n \, ds$$

show that ϕ must satisfy the condition

$$\iint_{\mathscr{S}} \phi \, d\sigma = 0 \tag{A}$$

[Physically, this restriction makes sense since $u_n = 0$ on $\mathscr{C}$ means there is no "flux" across $\mathscr{C}$; surely, then, the net "generation" $\iint_{\mathscr{S}} \phi \, d\sigma$ must be zero. In eigenfunction language (Comment 4, Section 5, of PART I), we note that zero is an eigenvalue of the operator $\mathbf{L} = \nabla^2$ over $\mathscr{S}$ plus the boundary condition $u_n = 0$ on $\mathscr{C}$, since u = constant = 1, for example, is a nontrivial solution of $\nabla^2 u = 0$, $u_n = 0$. Thus, (A) really states that ϕ must be orthogonal to the eigenfunction corresponding to the zero eigenvalue.] (b) Repeating the steps of part (a) for the boundary value problem governing G, we arrive at the analogous result

$$\int_{\mathscr{C}} G_n \, ds = \iint_{\mathscr{S}} \delta \, d\sigma \tag{B}$$

This is a contradiction, because G_n is required to be zero on $\mathscr{C}$ whereas the right-hand side is unity. Thus the ordinary Green's function does not exist! (c) Starting with the integration by parts formula

$$\iint_{\mathscr{S}} G\phi \, d\sigma = -\int_{\mathscr{C}} uG_n \, ds + \iint_{\mathscr{S}} u\nabla^2 G \, d\sigma \tag{C}$$

seek a "modified-generalized Green's function," along the lines of Exercise 4.14 of PART I. Show that if we require that $\nabla^2 G = \delta$ over $\mathscr{S}$, as usual, but that G_n be a (suitably chosen) constant over $\mathscr{C}$, then the contradiction in (B) is repaired, and the solution is

$$u = K + \iint_{\mathscr{S}} G\phi \, d\sigma \tag{D}$$

where K is an arbitrary constant. Verify directly that (D) does satisfy the required conditions $\nabla^2 u = \phi$ over $\mathscr{S}$ and $u_n = 0$ on $\mathscr{C}$.

6.10. The half-plane Poisson problem of Example 2 was posed with Dirichlet boundary conditions; $u = f(\xi)$ on $\eta = 0$, and hence $G = 0$ there. If, instead, we consider Neumann boundary conditions, $u_n = h(\xi)$ on $\eta = 0$, show that under certain additional restrictions the solution is

$$\begin{aligned} u(x, y) = &-\frac{1}{2\pi}\int_{-\infty}^{\infty} \ln\,[(\xi - x)^2 + y^2]h(\xi)\, d\xi \\ &+ \frac{1}{4\pi}\int_{-\infty}^{\infty}\int_{-\infty}^{\infty} \ln\,\{[(\xi - x)^2 + (\eta - y)^2][(\xi - x)^2 \\ &+ (\eta + y)^2]\}\phi(\xi, \eta)\, d\xi\, d\eta \end{aligned}$$

Integrating $\nabla^2 u = \phi$ over the region, as in Exercise 6.9, show that suitable "additional restrictions" are

$$\iint_{\mathscr{S}} \phi \, d\sigma = 0, \qquad \int_{-\infty}^{\infty} h \, d\xi = 0$$

and the boundary condition $u = O(r^{-\alpha})$ at infinity, where $\alpha > 0$.

6.11. For the limiting cases $\alpha \to 0$ and $\alpha \to \infty$, the mixed boundary condition of Example 3 reduces to Neumann and Dirichlet boundary conditions respectively. As a partial check on the correctness of the derived Green's function (6.65), verify that it does reduce to the appropriate simple forms for these two limiting cases. *Hint:* Note that

$$\int_0^{\infty} e^{-ax} f(x) \, dx \sim \frac{f(0)}{a} \text{ as } a \to \infty$$

6.12. Determine the Green's function for the problem $\nabla^2 u = \phi$ in the region bounded by the circle $\xi^2 + (\eta - h)^2 = R^2$ and the ξ axis, where u is prescribed on the circle and u_n on the ξ axis, as shown in the figure. *Hint:* First, place a suitable point mass at the image point a, to satisfy the boundary condition on the circle. But the boundary condition on the ξ axis is not satisfied, so add two suitable point masses at b and c. This, in turn, upsets the boundary condition on the circle, so add two more images at d, e. Continuing this process indefinitely, obtain G in the form of an infinite series.

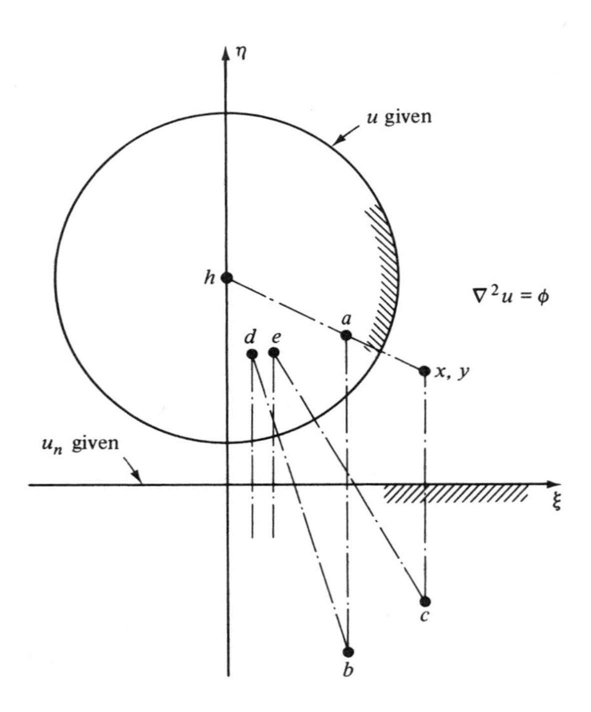

6.13. Determine the Green's function for the problem $\nabla^2 u = \phi$ in the annulus $1 < \xi^2 + \eta^2 < 4$, where u is given on both $\xi^2 + \eta^2 = 1$ and $\xi^2 + \eta^2 = 4$. *Hint:* Introduce an infinite sequence of images along the radial line through the point x, y.

6.14. Our image solution for the circular disk (Example 1) assumed Dirichlet boundary conditions. Can it be modified to accommodate Neumann boundary conditions?

7. GREEN'S FUNCTION METHOD FOR THE HELMHOLTZ OPERATOR

As our next application, we will consider the Helmholtz equation (1.6), $\nabla^2 u + k^2 u = \phi$. This will be discussed by means of two examples.

Example 1. *Vibrating Circular Membrane.* Suppose u is required to satisfy the Helmholtz equation throughout the interior of a disk of radius R, together with boundary values f, as shown in Fig. 7.1. Recall from Section 1 that k and u may be regarded as the frequency and amplitude of the displacement of a membrane which is driven by a periodic force distribution of amplitude ϕ.

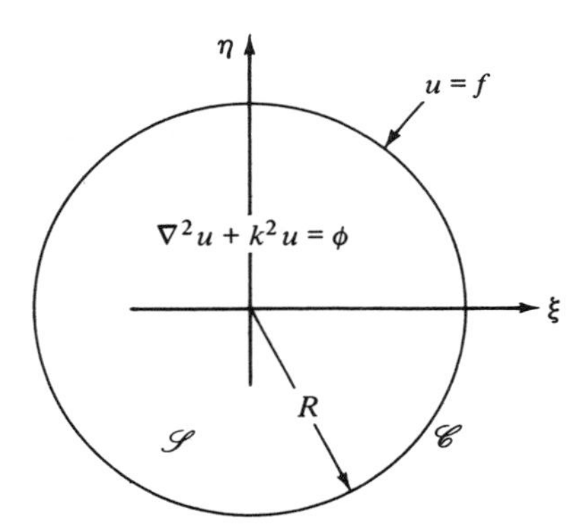

FIGURE 7.1 Summary of boundary value problem.

Proceeding exactly as with the Poisson equation in Section 6, the solution is again given by (6.3), where the Green's function now satisfies

$$\mathbf{L}^* G = G_{\xi\xi} + G_{\eta\eta} + k^2 G = \delta(\xi - x, \eta - y) \tag{7.1}$$

in $\mathscr{S}$, together with the condition $G = 0$ on $\mathscr{C}$. Recalling the principal solution (5.16), we seek G in the form

$$G(\xi, \eta; x, y) = U(\xi, \eta; x, y) + g(\xi, \eta; x, y) = \tfrac{1}{4} Y_0(kr) + g \tag{7.2}$$

where

$$r = \sqrt{(\xi - x)^2 + (\eta - y)^2} \tag{7.2a}$$

and

$$\mathbf{L}^* g = \nabla^2 g + k^2 g = g_{\xi\xi} + g_{\eta\eta} + k^2 g = 0 \tag{7.3}$$

throughout $\mathscr{S}$. Finally, $G = 0$ on $\mathscr{C}$ implies, from (7.2), that g must satisfy the boundary condition

$$g = -\tfrac{1}{4} Y_0(kr) \tag{7.4}$$

on $\mathscr{C}$. Changing to polar coordinates,

$$\begin{aligned} \xi &= \tilde{\rho} \cos \tilde{\vartheta} & \qquad x &= \rho \cos \vartheta \\ \eta &= \tilde{\rho} \sin \tilde{\vartheta} & \qquad y &= \rho \sin \vartheta \end{aligned}$$

and using separation of variables on the resulting version of (7.3), we obtain (Exercise 7.2)

$$g = \sum_{n=0}^{\infty} J_n(k\tilde{\rho})[a_n \cos n\tilde{\vartheta} + b_n \sin n\tilde{\vartheta}] \tag{7.5}$$

where the a_n's and b_n's are computed by application of the boundary condition (7.4),

$$-\tfrac{1}{4} Y_0[k\sqrt{R^2 + \rho^2 - 2R\rho \cos(\tilde{\vartheta} - \vartheta)}] = \sum_{n=0}^{\infty} J_n(kR)[a_n \cos n\tilde{\vartheta} + b_n \sin n\tilde{\vartheta}] \tag{7.6}$$

[The terms under the radical, in the argument of Y_0, result from changing to polar coordinates in (7.2a), with $\tilde{\rho}$ set equal to R.] Thus, the Green's function is given by (7.2) and (7.5) where

$$a_0 = -\frac{1}{8\pi J_0(kR)} \int_{-\pi}^{\pi} Y_0[k\sqrt{R^2 + \rho^2 - 2R\rho \cos(\tilde{\vartheta} - \vartheta)}]\, d\tilde{\vartheta}$$

and

$$\begin{aligned} a_n &= -\frac{1}{4\pi J_n(kR)} \int_{-\pi}^{\pi} Y_0[k\sqrt{R^2 + \rho^2 - 2R\rho \cos(\tilde{\vartheta} - \vartheta)}] \cos n\tilde{\vartheta}\, d\tilde{\vartheta} \\ b_n &= -\frac{1}{4\pi J_n(kR)} \int_{-\pi}^{\pi} Y_0[k\sqrt{R^2 + \rho^2 - 2R\rho \cos(\tilde{\vartheta} - \vartheta)}] \sin n\tilde{\vartheta}\, d\tilde{\vartheta} \end{aligned} \tag{7.7}$$

for $n \geq 1$.

Example 2. *Acoustic Radiation.* Now let us consider the two-dimensional acoustic radiation into the upper half-plane $y > 0$ due to a pulsating membrane stretched over $|x| < a$, as sketched in Fig. 7.2, where $\phi(x, y, t)$ is the velocity potential of the air. If the vertical displacement of the membrane is of the form $f(x) \sin \omega t$, then the vertical velocity of the membrane, and hence the vertical velocity ϕ_y of the adjacent air, is $\omega f(x) \cos \omega t$. Of course, the normal velocity ϕ_y is zero on the rigid wall, $|x| > a$ and $y = 0$. Thus, our boundary condition on ϕ is

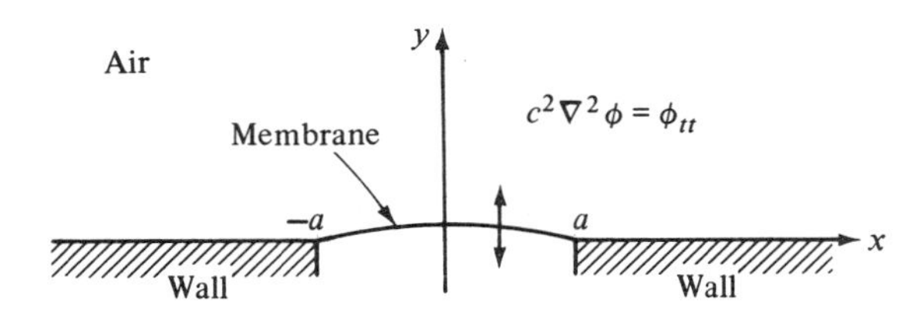

FIGURE 7.2 Acoustic radiation into $y > 0$.

$$\phi_y(x, 0, t) = \begin{cases} \omega f(x) \cos \omega t, & |x| < a \qquad (7.8a) \\ 0, & |x| > a \qquad (7.8b) \end{cases}$$

Actually, (7.8a) should be satisfied not on $y = 0$, but rather on the surface of the membrane, $y = f(x) \sin \omega t$. Our simplified version (7.8a) will, however, be an acceptable approximation if $f(x)$ is uniformly small over the interval.

It will be convenient to seek the solution in the form

$$\phi(x, y, t) = \mathscr{R}\{\psi(x, y)e^{i\omega t}\} \tag{7.9}$$

Inserting this into both the acoustic equation and (7.8) we arrive (see Exercise 7.3) at the following boundary value problem on $\psi(x, y)$, namely, the Helmholtz equation

$$\mathbf{L}\psi = \nabla^2\psi + k^2\psi = 0 \qquad \left(k \equiv \frac{\omega}{c}\right) \tag{7.10a}$$

in $y > 0$, together with the boundary condition

$$\psi_y(x, 0) = \begin{cases} \omega f(x), & |x| < a \\ 0, & |x| > a \end{cases} \tag{7.10b}$$

It is the boundary value problem (7.10) with which we shall be concerned.

To solve this problem by the method of Green's functions we choose "u" $= \psi$ and "v" $= G$ in (2.7). Since $\mathbf{L}\psi = 0$ in $\mathscr{S}$, we have

$$0 = \int_{-\infty}^{\infty} (-G\psi_\eta + \psi G_\eta)|_{\eta=0}\, d\xi + \iint_{\mathscr{S}} \psi \mathbf{L}^* G\, d\sigma \tag{7.11}$$

Now, whereas ψ_η is prescribed on $\eta = 0$, ψ is not. We therefore set $G_\eta = 0$ on $\eta = 0$, so that the requirements on G are

$$\mathbf{L}^* G = \nabla^2 G + k^2 G = G_{\xi\xi} + G_{\eta\eta} + k^2 G = \delta(\xi - x, \eta - y) \tag{7.12}$$

in $\eta > 0$, together with the boundary condition

$$G_\eta(\xi, 0; x, y) = 0 \tag{7.13}$$

With the help of (7.10b), (7.12), and (7.13), it follows from (7.11) that

$$\psi(x, y) = \omega \int_{-a}^{a} G(\xi, 0; x, y) f(\xi)\, d\xi \tag{7.14}$$

Finally, we must compute the Green's function. Recalling the principal solution (5.16) of the Helmholtz equation, we tentatively seek G in the form

$$G(\xi, \eta; x, y) = U(\xi, \eta; x, y) + g(\xi, \eta; x, y) = \tfrac{1}{4} Y_0(kr) + g \tag{7.15}$$

where

$$r = \sqrt{(\xi - x)^2 + (\eta - y)^2}$$

and

$$\mathbf{L}^* g = \nabla^2 g + k^2 g = 0 \tag{7.16}$$

throughout $\mathscr{S}$.

Before worrying about the boundary condition (7.13), let us examine the behavior of G at infinity. From the known asymptotic behavior of Y_0, we note that[20]

$$\frac{1}{4} Y_0(kr) \sim \frac{1}{4}\sqrt{\frac{2}{\pi kr}} \sin\left(kr - \frac{\pi}{4}\right) \qquad \text{as } r \to \infty \tag{7.17}$$

Combining this with our exp $(i\omega t)$ factor [recall (7.9)] and recalling that $k \equiv \omega/c$, we observe that the combination

$$\sin\left(kr - \frac{\pi}{4}\right) e^{i\omega t} = \frac{1}{2i}\{e^{ik(r+ct-\pi/4k)} - e^{-ik(r-ct-\pi/4k)}\} \tag{7.18}$$

contains both outgoing waves *and* incoming waves![21] But, *the presence of waves coming in from infinity is unacceptable on physical grounds;* this is the so-called **radiation condition.** To patch things up, let us discard our tentative choice of $Y_0(kr)/4$ as our principal solution, and replace it with one which does satisfy the radiation condition. Well, recall from Exercise (5.4) that two other principal solutions of the Helmholtz equation are $-iH_0^{(1)}(kr)/4$ and $iH_0^{(2)}(kr)/4$. Noting the asymptotic behavior,[22]

$$H_0^{(1)}(kr) \sim \sqrt{\frac{2}{\pi kr}}\, e^{i(kr-\pi/4)} \qquad \text{as } r \to \infty \tag{7.19}$$

[20] N. W. McLachlan, *Bessel Functions for Engineers*, 2nd ed., Oxford University Press, London, 1955, p. 197.

[21] The general form $F(r - ct)$ describes a wave moving radially outward at a constant speed c. Similarly, $F(r + ct)$ describes an inward-moving wave.

[22] McLachlan, *Bessel Functions for Engineers*, p. 198.

$$H_0^{(2)}(kr) \sim \sqrt{\frac{2}{\pi kr}}\, e^{-i(kr-\pi/4)} \qquad \text{as } r \to \infty \tag{7.20}$$

we see that the principal solution $iH_0^{(2)}(kr)/4$ does in fact satisfy the radiation condition; i.e., introducing our $\exp(i\omega t) = \exp(ikct)$ factor we have, from (7.20),

$$\frac{i}{4} H_0^{(2)}(kr) e^{i\omega t} \sim \frac{i}{4}\sqrt{\frac{2}{\pi kr}}\, e^{-ik(r-ct-\pi/4k)} \tag{7.21}$$

which is purely an *outgoing* wave. (Note that its amplitude diminishes as it moves outward because of the $r^{-1/2}$ factor.) Thus, choosing $iH_0^{(2)}(kr)/4$ as our principal solution U, in place of $Y_0(kr)/4$, it now remains to determine a g which permits satisfaction of the boundary condition (7.13) and, at the same time, satisfies the radiation condition itself. This is easily accomplished, by the method of images, by choosing $g = U(\xi, \eta; x, -y)$. Thus,

$$\begin{aligned} G(\xi, \eta; x, y) &= \frac{i}{4} H_0^{(2)}[k\sqrt{(\xi - x)^2 + (\eta - y)^2}] \\ &\quad + \frac{i}{4} H_0^{(2)}[k\sqrt{(\xi - x)^2 + (\eta + y)^2}] \end{aligned} \tag{7.22}$$

and the solution, according to (7.14), is therefore

$$\psi(x, y) = \frac{i\omega}{2}\int_{-a}^{a} f(\xi) H_0^{(2)}[k\sqrt{(\xi - x)^2 + y^2}]\, d\xi \tag{7.23}$$

COMMENT 1. Notice that we didn't *have* to discard the $Y_0(kr)/4$ principal solution. We could have kept it, and required g to contain an incoming wave part that would exactly cancel the incoming wave contained in $Y_0(kr)/4$. Noting that the solution $J_0(kr)$ of (7.16) behaves like[23]

$$J_0(kr) \sim \sqrt{\frac{2}{\pi kr}} \cos\left(kr - \frac{\pi}{4}\right) \qquad \text{as } r \to \infty \tag{7.24}$$

we are led to the combination

$$g = \frac{i}{4} J_0(kr) + \frac{i}{4} H_0^{(2)}[k\sqrt{(\xi - x)^2 + (\eta + y)^2}] \tag{7.25}$$

That is, the J_0 term in (7.25) serves to cancel the incoming wave portion of the principal solution, since

$$\begin{aligned} \frac{1}{4} Y_0(kr) + \frac{i}{4} J_0(kr) &\sim \frac{1}{4}\sqrt{\frac{2}{\pi kr}}\left\{\sin\left(kr - \frac{\pi}{4}\right) + i\cos\left(kr - \frac{\pi}{4}\right)\right\} \\ &= \frac{i}{4}\sqrt{\frac{2}{\pi kr}}\, e^{-i(kr-\pi/4)} \end{aligned} \tag{7.26}$$

[23] McLachlan, *Bessel Functions for Engineers*, p. 191.

and when our $\exp(i\omega t) = \exp(ikct)$ factor is introduced this becomes $(i/\sqrt{8\pi kr}) \exp[-ik(r - ct - \pi/4k)]$, which is purely *outgoing*. As for the boundary condition $G_\eta(\xi, 0; x, y) = 0$, we note that this is taken care of since the $H_0^{(2)}$ term in (7.25) is the image of both the J_0 and Y_0 terms. Of course, the resulting Green's function is exactly the same as before [i.e., (7.22)], because of the identity $Y_0(kr) + iJ_0(kr) = iH_0^{(2)}(kr)$.

COMMENT 2. For our region $\mathscr{S}$ to be closed, we consider its boundary to consist of the ξ axis together with a large semicircle. The integral over the semicircle was tacitly omitted in (7.11), and we leave it for the reader to verify a posteriori that it does in fact tend to zero as the radius of the semicircle is allowed to become infinite (Exercise 7.4).

COMMENT 3. Finally, it is interesting to revisit Example 1 in connection with the radiation condition. Since the region was finite, there was no need to worry about waves coming in from infinity. Actually, if we ignore for the moment the fact that the domain is bounded, and examine the solution at infinity we will, in fact, discover both incoming and outgoing waves. The same will be true within our domain $\mathscr{S}$. The superposition of these incoming and outgoing waves is such that a given wave in our domain $\mathscr{S}$ appears to be continually reflected from the center of the disk and its outer edge, in turn.

EXERCISES

7.1. Instead of splitting the Green's function, of Example 1, according to (7.2), solve for G directly from the governing equation (7.1) and boundany condition $G = 0$ on $\tilde{\rho} = R$. *Hint:* To avoid the delta function in (7.1), divide the disk into two parts, $0 \leq \tilde{\rho} < \rho$ and $\rho < \tilde{\rho} < R$, in each of which G satisfies the homogeneous version of (7.1). The unknown coefficients that arise are determined by requiring boundedness of the "inner" solution, applying the condition $G = 0$ on $\tilde{\rho} = R$ to the "outer" solution, and blending the two solutions along $\tilde{\rho} = \rho$ by integrating (7.1) over the neighborhood of the point of action of the delta function. (This line of approach is basically the same as the one used so often in PART I in connection with *ordinary* differential equations. Its unwieldiness here explains why we tend to favor the principal solution approach for partial differential equations.)

7.2. Express (7.3) in terms of the polar coordinates $\tilde{\rho}$ and $\tilde{\vartheta}$. Then, using separation of variables, derive the expression (7.5) for g.

7.3. Verify that (7.10a) is indeed implied by the form (7.9) and the acoustic equation on ϕ. *Hint:* The key point is that $\mathbf{L}\mathscr{R}\{\ \} = \mathscr{R}\mathbf{L}\{\ \}$ whenever $\mathbf{L}$ is a *linear* operator; recall equation (1.2) of PART I.

7.4. Verify, a posteriori, that the integral over the large semicircle does vanish as the radius tends to infinity, as stated in Comment 2.

8. GREEN'S FUNCTION METHOD FOR THE DIFFUSION OPERATOR

We turn our attention now to the one-dimensional diffusion equation (1.9), $\mathbf{L}u = \kappa u_t - u_{xx} = \phi$. For definiteness, we will consider it within the physical context of heat conduction, as outlined in Section 1.

Our starting point is equation (2.9). Choosing "v" to be the Green's function G, and requiring that

$$\mathbf{L}^*G = -\kappa G_\tau - G_{\xi\xi} = \delta(\xi - x, \tau - t) \tag{8.1}$$

(2.9) reduces to

$$\iint_{\mathscr{S}} G\phi \, d\sigma = \int_{\mathscr{C}} [uG_\xi - Gu_\xi)\mathbf{i} + \kappa uG\mathbf{j}]\cdot\mathbf{n} \, ds + u(x, t) \tag{8.2}$$

Example 1. *Semi-infinite Rod.* Consider our heat conducting rod to be semi-infinite in length, with its initial temperature distribution $u(x, 0)$ and end temperature $u(0, t)$ prescribed to be $f(x)$ and $h(t)$, respectively. Now, in seeking $u(x, t)$ from (8.2) let us consider a time duration T; the value of T will be seen to be immaterial, as long as it exceeds any particular value of t at which the solution is sought. The problem is summarized in Fig. 8.1. Clearly, (8.2) reduces to

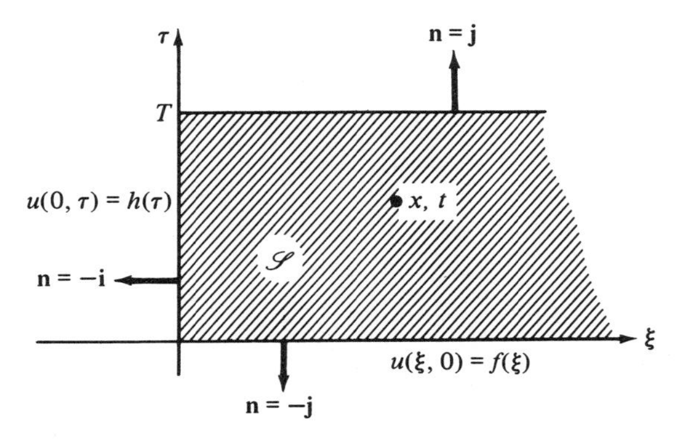

FIGURE 8.1 Problem summary, Example 1.

$$\iint_{\mathscr{S}} G\phi \, d\sigma = \int_0^T (-uG_\xi + Gu_\xi)|_{\xi=0} \, d\tau - \int_0^\infty (\kappa uG)|_{\tau=0} \, d\xi$$
$$+ \int_0^\infty (\kappa uG)|_{\tau=T} \, d\xi + u(x, t) \tag{8.3}$$

Now, u_ξ is not prescribed on $\xi = 0$, and u is not prescribed on $\tau = T$. To remove these unwelcome terms in (8.3), let us require that $G = 0$ along both $\xi = 0$ and $\tau = T$. Now, since $G(\xi, T; x, t) = 0$ where T is an *arbitrary* time, greater than t, it follows that $G(\xi, \tau; x, t) = 0$ for *all* $\tau > t$. Provided that we can in fact find G, the solution follows from (8.3) as

$$u(x, t) = \int_0^t h(\tau)G_\xi(0, \tau; x, t) \, d\tau + \int_0^\infty \kappa f(\xi)G(\xi, 0; x, t) \, d\xi$$
$$+ \int_0^t \int_0^\infty G(\xi, \tau; x, t)\phi(\xi, \tau) \, d\xi \, d\tau \tag{8.4}$$

where the upper limit in the first integral has been changed from T to t since G, and hence G_ξ, is identically zero for all $\tau > t$.

We seek G in the form

$$G(\xi, \tau; x, t) = U(\xi, \tau; x, t) + g(\xi, \tau; x, t) \tag{8.5}$$

where, from (5.23),

$$U = \frac{H(t - \tau)e^{-\kappa(\xi - x)^2/4(t-\tau)}}{\sqrt{4\pi\kappa(t - \tau)}} \tag{8.6}$$

Recall from our discussion in Section 5 that this principal solution describes the diffusion of a unit heat pulse placed at x, t in a doubly infinite rod—but "backward" in time. Clearly, we can construct g by a simple image system, specifically, a *negative* unit heat pulse at $-x, t$. That is, $g = -U(\xi, \tau; -x, t)$ and hence

$$G(\xi, \tau; x, t) = U(\xi, \tau; x, t) - U(\xi, \tau; -x, t) \tag{8.7}$$

COMMENT 1. For the sake of illustration, consider the case where $h(t) = \phi(x, t) = 0$ and $f(x) = \delta(x - a)$; that is, a heat source is applied at $x = a$ and $t = 0$, with the left end "quenched" for all $t > 0$. From (8.4), (8.6), and (8.7), the solution is simply

$$u(x, t) = \sqrt{\frac{\kappa}{4\pi t}} \, [e^{-\kappa(x-a)^2/4t} - e^{-\kappa(x+a)^2/4t}] \tag{8.8}$$

as sketched in Fig. 8.2. This illustrates a striking and well-known feature of the diffusion equation; for $t > 0$, but *arbitrarily* small, $u(x, t) \neq 0$ for all $x > 0$. That is, the heat has diffused with an *infinite velocity*! Since propagation of energy at an infinite velocity is impossible, it appears that the diffusion equation can only be correct after a very small amount of time has elapsed. Insofar as practical calculations are concerned, this deficiency is rather minor and is generally ignored completely. It is interesting, however, that to patch things up the diffusion equation should be modified[24] to read

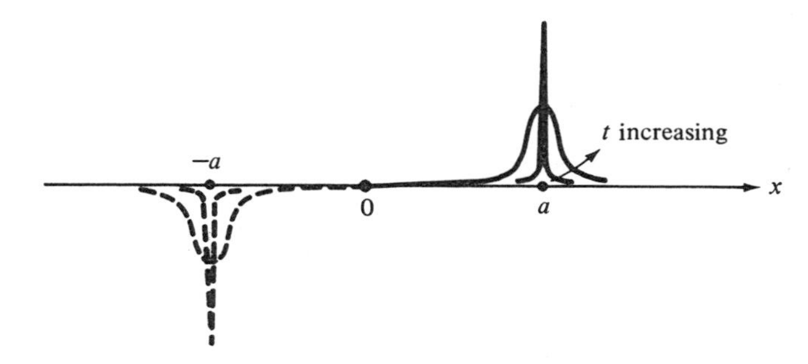

FIGURE 8.2 Temperature variation given by Eq. (8.8).

$$u_{xx} = \kappa u_t + \frac{1}{c^2} u_{tt} \tag{8.9}$$

where c is the acoustic velocity in the material. Strictly speaking, (8.9) is *hyperbolic* for all x and t, due to the addition of the u_{tt}/c^2 term. However, observe that the $1/c^2$ factor is *extremely small*. Thus, except for the initial instants when u_{tt} is large, the last term becomes negligible, and the character of (8.9) becomes dominated by the diffusion terms. (See Exercise 8.5.)

COMMENT 2. Let us use our Example 1 to illustrate a general scheme proposed by B. A. Boley[25] for the iterative calculation of Green's functions. We return to our starting point, equation (2.9), and set

$$\begin{aligned} \text{“}v\text{”} &= G(\xi', \tau'; x, t) \\ \text{“}u\text{”} &= g^*(\xi', \tau'; \xi, \tau) \end{aligned} \tag{8.10}$$

[By g^* we mean the regular part of G^*; that is, $G^* = U^* + g^*$. At this point it may be helpful to re-read Comment 2 on Example 1 of Section 6. Recalling the general relation (6.30), it is clear that the same relationship must hold between the singular parts U^* and U, and between the regular parts g^* and g. In graphic terms, our U^* and U correspond to the "forward" and "backward"

[24] P. M. Morse and H. Feshbach, *Methods of Theoretical Physics,* Part I, McGraw-Hill Book Company, New York, 1953, pp. 865–869.

[25] B. A. Boley, "A Method for the Construction of Green's Functions," *Quarterly of Applied Mathematics,* Vol. XIV, No. 3, October, 1956, pp. 249–257.

diffusion, respectively, of a unit heat pulse in a doubly infinite rod.] With "v" and "u" chosen according to (8.10), (2.9) becomes

$$0 = \int_0^t (-g^* G_{\xi'} + G g^*_{\xi'})|_{\xi'=0}\, d\tau' + \int_0^\infty (\kappa g^* G)|_{\tau'=T}\, d\xi'$$
$$- \int_0^\infty (\kappa g^* G)|_{\tau'=0}\, d\xi' + \iint_{\mathscr{S}} g^*(\xi', \tau'; \xi, \tau)\delta(\xi' - x, \tau' - t)\, d\xi'\, d\tau' \tag{8.11}$$

But $G = 0$ on both $\xi' = 0$ and $\tau' = T$, and $g^* = 0$ on $\tau' = 0$ (since both G^* and $U^* = 0$ there; see Exercise 8.1), so that

$$0 = -\int_0^t g^*(0, \tau'; \xi, \tau) G_{\xi'}(0, \tau'; x, t)\, d\tau' + g^*(x, t; \xi, \tau)$$

or

$$g(\xi, \tau; x, t) = \int_0^t g(\xi, \tau; 0, \tau') G_{\xi'}(0, \tau'; x, t)\, d\tau' \tag{8.12}$$

Finally, with $g = -U$ on $\xi' = 0$ since $G = 0$ there, and $G_{\xi'}$ replaced by $U_{\xi'} + g_{\xi'}$, we obtain the integral equation

$$g(\xi, \tau; x, t) = -\int_0^t U(\xi, \tau; 0, \tau') U_{\xi'}(0, \tau'; x, t)\, d\tau'$$
$$-\int_0^t U(\xi, \tau; 0, \tau') g_{\xi'}(0, \tau'; x, t)\, d\tau' \tag{8.13}$$

on the unknown g. To solve, Boley suggests the iterative scheme

$$g^{(n+1)} = -\int_0^t U U_{\xi'}\, d\tau' - \int_0^t U g^{(n)}_{\xi'}\, d\tau' \tag{8.14}$$

Starting with $g^{(0)} = 0$, the recursion formula (8.14) produces (Exercise 8.6)

$$g^{(1)}(\xi, \tau; x, t) = -\frac{1}{2} U(\xi, \tau; -x, t)$$
$$g^{(2)}(\xi, \tau; x, t) = -\frac{7}{8} U(\xi, \tau; -x, t)$$
$$\vdots \qquad\qquad \vdots$$
$$g^{(i)}(\xi, \tau; x, t) = -(1 - 2^{1-2^i}) U(\xi, \tau; -x, t)$$
$$\to -U(\xi, \tau; -x, t) \qquad \text{as } i \to \infty \tag{8.15}$$

which agrees which the exact result as stated in (8.7). Although the convergence is very rapid, it is only fair to mention that in more complicated applications the successive integrations required in the iteration may be prohibitively difficult.

EXERCISES

8.1. Verify our claim, in Comment 2, that $G^* = 0$ on $\tau' = 0$. *Hint:* Recall that the adjoint boundary conditions are implied by the expression $(\mathbf{L}G^*, G) = (G^*, \mathbf{L}^* G)$. Thus the boundary conditions on G^* must be such that the boundary integral in (2.9),

$$\int_{\mathscr{C}} [(G^* G_\xi - G G_\xi^*)\mathbf{i} + \kappa G^* G\mathbf{j}] \cdot \mathbf{n}\, ds$$

is zero. Recalling that $G = 0$ on the edges $\xi = 0$ and $\tau = T$, show that we must have $G^* = 0$ on $\xi = 0$ and $\tau = 0$.

8.2. Verify that our solution (8.4) does in fact satisfy the required initial condition, $u(x, t) \to f(x)$ as $t \to 0$. *Hint:* Recall the δ-sequence of Exercise 3.1(c) in PART I.

8.3. Deduce, by the method of images, the Green's function for a *finite* rod with end conditions $u(0, t) = h(t), u_x(\ell, t) = 0$, and an initial condition $u(x, 0) = f(x)$.

8.4. Find the Green's function, and hence the solution, for the case of a semi-infinite rod with initial temperature $u(x, 0) = 0$, subjected to a *mixed* boundary condition $\partial u/\partial n + \beta u = -\partial u/\partial x + \beta u = h(t)$ at the left end $(x = 0)$, where β is a constant. *Hint:* Instead of using Moran's technique, discussed in Section 6, it is more direct to use the results of that example (see Fig. 6.8) to help us anticipate a Green's function of the form

$$G(\xi, \tau; x, t) = U(\xi, \tau; x, t) + AU(\xi, \tau; -x, t) + \int_{-\infty}^{-x} a(\zeta)U(\xi, \tau; \zeta, t)\, d\zeta$$

where A and $a(\zeta)$ are to be suitably chosen. Show that the conditions required of G are all satisfied if we have $A = 1, a(-x) + 2\beta = 0$, and $a'(\zeta) - \beta a(\zeta) = 0$ over $-\infty < \zeta < -x$, so that

$$G = \frac{H(t-\tau)}{\sqrt{4\pi\kappa(t-\tau)}} \Big\{ e^{-\kappa(\xi-x)^2/4(t-\tau)} + e^{-\kappa(\xi+x)^2/4(t-\tau)} - 2\beta e^{\beta x} \int_{-\infty}^{-x} e^{\beta\zeta - \kappa(\zeta-\xi)^2/4(t-\tau)}\, d\zeta \Big\}$$

and

$$u(x, t) = \int_0^t h(\tau) G(0, \tau; x, t)\, d\tau$$

(The *uniqueness* of both G and u are established on pages 82–83 of A. Sommerfeld's *Partial Differential Equations in Physics*, Academic Press, Inc., New York, 1949.)

8.5. The principal solution corresponding to the differential operator $\mathbf{L} = \partial^2/\partial\xi^2 - \kappa\partial/\partial\tau - (1/c^2)\partial^2/\partial\tau^2$ of equation (8.9) satisfies the equation

$$\mathbf{L}^* U = U_{\xi\xi} + \kappa U_\tau - \frac{1}{c^2} U_{\tau\tau} = \delta(\xi - x, \tau - t).$$

(The result

$$U(\xi, \tau; x, t) = -\frac{c}{2} H[c(t-\tau) - |\xi - x|] e^{-\kappa c^2(t-\tau)/2} \times J_0\left[\frac{\kappa c}{2}\sqrt{(\xi - x)^2 - c^2(t-\tau)^2}\right] \tag{A}$$

is derived in Morse and Feshbach, *Methods of Theoretical Physics*, Part I, pp. 865–869). Show that for the two limiting cases, $\kappa \to 0$ and $c \to \infty$, (A) does reduce to the principal solutions that we derived in Section 5 for the wave and diffusion differential operators, namely,

$$U(\xi, \tau; x, t) = -\frac{c}{2} H[c(t-\tau) - |\xi - x|]$$

corresponding to

$$\mathbf{L}^* U = U_{\xi\xi} - \frac{1}{c^2} U_{\tau\tau} = \delta(\xi - x, \tau - t),$$

and

$$U(\xi, \tau; x, t) = -\frac{H(t-\tau)e^{-\kappa(\xi - x)^2/4(t-\tau)}}{\sqrt{4\pi\kappa(t-\tau)}}$$

corresponding to $\mathbf{L}^* U = \kappa U_\tau + U_{\xi\xi} = \delta(\xi - x, \tau - t)$. *Hint:* $J_0(0) = 1$, $J_0(ix) = I_0(x)$ where I_0 is the *modified Bessel function of the first kind and order zero*, and $I_0(x) \sim e^x/\sqrt{2\pi x}$ as $x \to \infty$. (McLachlan's *Bessel Functions for Engineers*, pp. 190, 200, and 201.)

9. GREEN'S FUNCTION METHOD FOR THE WAVE OPERATOR

Finally, we consider the one-dimensional wave equation (1.10), $\mathbf{L}u = c^2 u_{xx} - u_{tt} = \phi$. This time our starting point is equation (2.10), with "v" chosen to be our Green's function G. Requiring that

$$\mathbf{L}^* G = c^2 G_{\xi\xi} - G_{\tau\tau} = \delta(\xi - x, \tau - t) \tag{9.1}$$

we have

$$\iint_{\mathscr{S}} G\phi \, d\sigma = \int_{\mathscr{C}} [c^2(Gu_\xi - uG_\xi)\mathbf{i} - (Gu_\tau - uG_\tau)\mathbf{j}] \cdot \mathbf{n} \, ds + u(x, t) \tag{9.2}$$

Example 1. *Doubly-infinite String.* Consider a doubly-infinite string, set in motion by means of an initial displacement $u(x, 0) = f(x)$ and an initial velocity $u_t(x, 0) = h(x)$. As with the heat equation, we consider a time duration T which exceeds t but is otherwise arbitrary; see Fig. 9.1. With the forcing function $\phi \equiv 0$, for simplicity, (9.2) reduces to

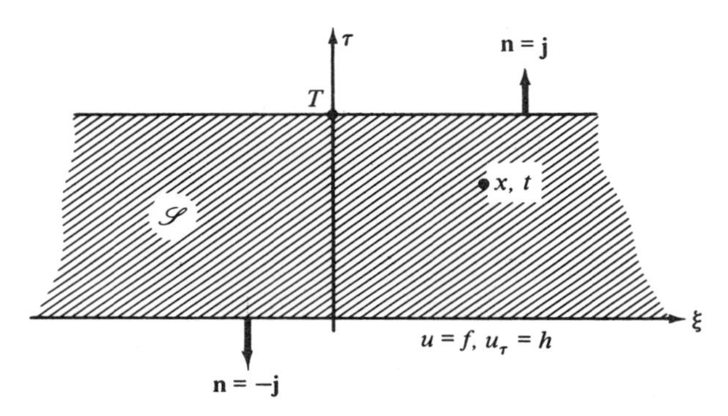

FIGURE 9.1 Problem summary, Example 1.

$$0 = \int_{-\infty}^{\infty} (Gu_\tau - uG_\tau)|_{\tau=0}\, d\xi - \int_{-\infty}^{\infty} (Gu_\tau - uG_\tau)|_{\tau=T}\, d\xi + u(x, t) \tag{9.3}$$

Now, whereas u and u_τ are known on $\tau = 0$ they are not known on $\tau = T$. To remove these unwelcome terms from (9.3) we therefore subject G to the conditions $G = G_\tau = 0$ on $\tau = T$; or, since T is arbitrarily greater than t, we require simply that $G \equiv 0$ for all $\tau > t$. We therefore have

$$u(x, t) = \int_{-\infty}^{\infty} [f(\xi)G_\tau(\xi, 0; x, t) - h(\xi)G(\xi, 0; x, t)]\, d\xi \tag{9.4}$$

Finally, we observe that our Green's function is simply the "backward running" principal solution, given by (5.26) and Fig. 5.2. Using Heaviside function notation, we may express

$$G(\xi, \tau; x, t) = -\frac{H(t-\tau)}{2c}\{H[x + c(t-\tau) - \xi] - H[x - c(t-\tau) - \xi]\} \tag{9.5}$$

Differentiating, with the help of equation (3.26) of PART I,

$$G_\tau(\xi, \tau; x, t) = \frac{H(t-\tau)}{2}\{\delta[x + c(t-\tau) - \xi] + \delta[x - c(t-\tau) - \xi]\} \tag{9.6}$$

and inserting these in (9.4), with $\tau = 0$, we finally obtain the well-known **D'Alembert formula**

$$u(x, t) = \frac{f(x + ct) + f(x - ct)}{2} - \frac{1}{2c}\int_{x-ct}^{x+ct} h(\xi)\, d\xi \tag{9.7}$$

EXERCISES

9.1. Show that the adjoint boundary conditions for our wave operator are $G^* = G^*_\tau = 0$ on $\tau = 0$, whereas $G = G_\tau = 0$ on $t = T$. Thus the operator $\mathscr{G}$ is not self-adjoint even though $\mathbf{L} = \mathbf{L}^*$.

9.2. Obtain the Green's function for a *semi*-infinite string subjected to the conditions $u(x, 0) = u_t(x, 0) = 0$ over $0 < x < \infty$, and $u(0, t) = \epsilon \sin \omega t$ for all $t \geq 0$.

9.3. Obtain the additional term, reduced to its simplest form, which is present in the D'Alembert formula in the event that the forcing function $\phi(x, t)$ is *not* identically zero.

10. THE EIGENFUNCTION METHOD

In Section 5 of PART I we pointed out that there are basically two alternative lines of approach to the problem $\mathbf{L}u = \phi$: the inverse operator or Green's function method, and the eigenfunction method. Whereas eigenfunction expansions are certainly of great importance, our present concern is with *Green's functions*; our discussion of the eigenfunction method, both in PART I and here, is therefore quite brief, and concerned largely with the interrelationship between the two methods.

To illustrate the extension of the eigenfunction method (outlined in PART I) to partial differential equations let us consider the following application.

Example 1. *Poisson Equation for a Rectangle.* We consider the boundary value problem

$$\mathbf{L}u = \nabla^2 u = \phi \tag{10.1}$$

over the rectangular region shown in Fig. 10.1, with the boundary values $u = f$ prescribed on $\mathscr{C}$. Recalling from PART I that *homogeneous* boundary conditions are required for the application of the eigenfunction method, we see that we cannot proceed unless $f = 0$, so that the only inhomogeneity is in the differential equation.

Of course, it may be possible to "homogenize" the boundary conditions by a suitable change of variables, as discussed in Part I. That is, suppose we

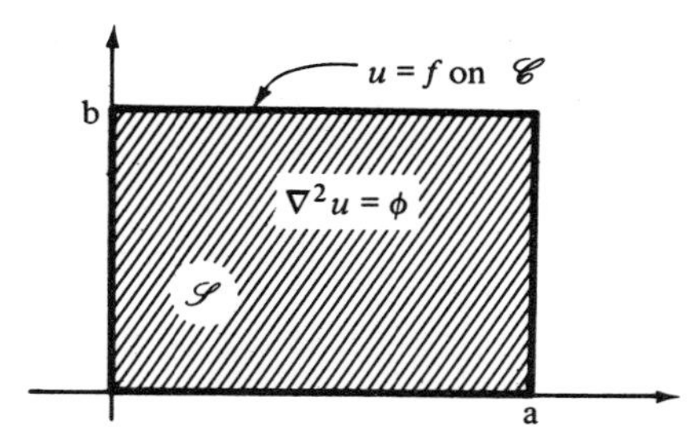

FIGURE 10.1 Boundary value problem, Example 1.

can construct (hopefully, by inspection) a function Φ which is "suitably well behaved"[26] over the region $\mathscr{S}$, and takes on the prescribed values f on $\mathscr{C}$. Then, changing the dependent variable from u to v according to $v = u - \Phi$, we have $\mathbf{L}u = \mathbf{L}(v + \Phi) = \mathbf{L}v + \mathbf{L}\Phi = \phi$, so that our boundary value problem becomes

$$\mathbf{L}v = \{\phi - \mathbf{L}\Phi\} \tag{10.2}$$

together with the *homogeneous* boundary condition $v = 0$ on $\mathscr{C}$. Construction of Φ may, however, be difficult—depending upon the boundary values f and the shape of the region. The simplest case occurs when f is a constant; then we simply choose $\Phi = f$, independent of the shape of $\mathscr{S}$.

In any event, the corresponding boundary value problem on the Green's function *will* contain homogeneous boundary conditions,

$$\begin{aligned} \mathbf{L}^*G &= \nabla^2 G = \delta(\xi - x, \eta - y) \\ G &= 0 \text{ on } \mathscr{C} \end{aligned} \tag{10.3}$$

and the eigenfunction method can be used to determine G.

Now, the eigenvalue problem corresponding to (10.3) is

$$\nabla^2 u + \lambda u = 0, \qquad u = 0 \text{ on } \mathscr{C} \tag{10.4}$$

Solving this by "separation of variables," we seek

$$u(\xi, \eta) = X(\xi)Y(\eta) \tag{10.5}$$

Inserting this into (10.4), and dividing through by XY we have

$$\frac{X''(\xi)}{X(\xi)} = -\frac{Y''(\eta)}{Y(\eta)} - \lambda \tag{10.6}$$

Since the left-hand side is a function of ξ alone, and the right-hand side is a function of η alone, we observe that (10.6) can hold for all ξ's and η's in $\mathscr{S}$

[26] Looking ahead to (10.2), we see that it must at least be sufficiently differentiable over $\mathscr{S}$ for $\mathbf{L}\Phi$ to exist.

only if both sides are in fact constant, say $-k^2$. As for the boundary conditions on X and Y, the conditions $u(0, \eta) = u(\mathrm{a}, \eta) = u(\xi, 0) = u(\xi, \mathrm{b}) = 0$ imply that $X(0) = X(\mathrm{a}) = Y(0) = Y(\mathrm{b}) = 0$. Thus,

$$X'' + k^2 X = 0; \qquad X(0) = X(\mathrm{a}) = 0 \tag{10.7a}$$

$$Y'' + (\lambda - k^2)Y = 0; \qquad Y(0) = Y(\mathrm{b}) = 0 \tag{10.7b}$$

From (10.7a),

$$X(\xi) = A \sin k\xi + B \cos k\xi \tag{10.8a}$$

$$X(0) = 0 = B \tag{10.8b}$$

$$X(\mathrm{a}) = 0 = A \sin k\mathrm{a} \tag{10.8c}$$

To satisfy (10.8c), either A and/or $\sin k\mathrm{a}$ must be zero; $A = 0$ is unacceptable since it leads to the trivial solution $X(\xi) = 0$ and hence $u(\xi, \eta) = 0$, whereas eigenfunctions are, by definition, *non*trivial solutions of (10.4). To have a *non*-trivial solution, the "separation constant" k must be chosen such that $k\mathrm{a}$ coincides with a zero of the sine function, i.e., $k\mathrm{a} = m\pi$, where $m = 1, 2, \ldots$. The constant A remains arbitrary, and

$$X(\xi) = A \sin \frac{m\pi\xi}{\mathrm{a}} \tag{10.9}$$

Inserting $k = m\pi/\mathrm{a}$ into (10.7b), and proceeding as above, we have

$$Y(\eta) = C \sin \sqrt{\lambda - \left(\frac{m\pi}{\mathrm{a}}\right)^2}\,\eta + D \cos \sqrt{\lambda - \left(\frac{m\pi}{\mathrm{a}}\right)^2}\,\eta \tag{10.10a}$$

$$Y(0) = 0 = D \tag{10.10b}$$

$$Y(\mathrm{b}) = 0 = C \sin \sqrt{\lambda - \left(\frac{m\pi}{\mathrm{a}}\right)^2}\,\mathrm{b} \tag{10.10c}$$

so that, for nontrivial solutions, we need

$$\sqrt{\lambda - \left(\frac{m\pi}{\mathrm{a}}\right)^2}\,\mathrm{b} = n\pi, \qquad \text{for } n = 1, 2, \ldots$$

or

$$\lambda = \pi^2\left(\frac{m^2}{\mathrm{a}^2} + \frac{n^2}{\mathrm{b}^2}\right) \equiv \text{“}\lambda_{mn}\text{”} \tag{10.11}$$

and

$$Y(\eta) = C \sin \frac{n\pi\eta}{\mathrm{b}} \tag{10.12}$$

Thus, the eigenvalues of (10.4) are given by (10.11), and the corresponding eigenfunctions are

$$\sin \frac{m\pi\xi}{\mathrm{a}} \sin \frac{n\pi\eta}{\mathrm{b}} \equiv \text{"}\phi_{mn}(\xi, \eta)\text{"} \tag{10.13}$$

Next, we expand the quantities in (10.3) in terms of these eigenfunctions:

$$G(\xi, \eta; x, y) = \sum_m \sum_n a_{mn}(x, y)\phi_{mn}(\xi, \eta) \tag{10.14}$$

$$\delta(\xi - x, \eta - y) = \sum_m \sum_n c_{mn}(x, y)\phi_{mn}(\xi, \eta) \tag{10.15}$$

Whereas the a_{mn} Fourier coefficients are the unknowns, the c_{mn}'s are given by [cf. (5.12) of PART I]

$$c_{mn} = \frac{(\delta, \phi_{mn})}{(\phi_{mn}, \phi_{mn})} \tag{10.16}$$

where we define the "two-dimensional" inner product by

$$(f, g) = \iint_{\mathscr{S}} fg\, d\sigma \tag{10.17}$$

$$= \int_0^{\mathrm{b}} \int_0^{\mathrm{a}} fg\, d\xi\, d\eta \tag{10.18}$$

Accordingly,

$$(\delta(\xi - x, \eta - y), \phi_{mn}(\xi, \eta)) = \phi_{mn}(x, y) \tag{10.19}$$

so that

$$c_{mn} = \frac{\phi_{mn}(x, y)}{(\phi_{mn}, \phi_{mn})} \tag{10.20}$$

Now, inserting (10.14) and (10.15) into (10.3), and noting from (10.4) that $\nabla^2\phi_{mn} = -\lambda_{mn}\phi_{mn}$, we have

$$-\sum_m \sum_n \lambda_{mn} a_{mn}(x, y)\phi_{mn}(\xi, \eta) = \sum_m \sum_n \frac{\phi_{mn}(x, y)}{(\phi_{mn}, \phi_{mn})} \phi_{mn}(\xi, \eta) \tag{10.21}$$

so that, equating coefficients of $\phi_{mn}(\xi, \eta)$,

$$a_{mn}(x, y) = -\frac{\phi_{mn}(x, y)}{\lambda_{mn}(\phi_{mn}, \phi_{mn})} \tag{10.22}$$

and [cf. (5.27) of PART I],

$$G(\xi, \eta; x, y) = -\sum_m \sum_n \frac{\phi_{mn}(x, y)\phi_{mn}(\xi, \eta)}{\lambda_{mn}(\phi_{mn}, \phi_{mn})} \tag{10.23}$$

With ϕ_{mn} given by (10.13), λ_{mn} by (10.11), and

$$(\phi_{mn}, \phi_{mn}) = \int_0^{\mathrm{b}} \int_0^{\mathrm{a}} \sin^2 \frac{m\pi\xi}{\mathrm{a}} \sin^2 \frac{n\pi\eta}{\mathrm{b}}\, d\xi\, d\eta$$

$$= \left\{\int_0^{\mathrm{b}} \sin^2 \frac{n\pi\eta}{\mathrm{b}}\, d\eta\right\} \left\{\int_0^{\mathrm{a}} \sin^2 \frac{m\pi\xi}{\mathrm{a}}\, d\xi\right\}$$

$$= \left\{\frac{\mathrm{b}}{2}\right\} \left\{\frac{\mathrm{a}}{2}\right\} = \frac{\mathrm{ab}}{4} \tag{10.24}$$

we have, finally,

$$G(\xi, \eta; x, y) = -4\,\mathrm{ab} \sum_m \sum_n \frac{\sin\left(\frac{m\pi x}{\mathrm{a}}\right) \sin\left(\frac{n\pi y}{\mathrm{b}}\right) \sin\left(\frac{m\pi\xi}{\mathrm{a}}\right) \sin\left(\frac{n\pi\eta}{\mathrm{b}}\right)}{\pi^2(m^2\mathrm{b}^2 + n^2\mathrm{a}^2)} \tag{10.25}$$

COMMENT 1. Our eigenfunction procedure has been basically identical to that developed in PART I. Whereas we justified our expansions in PART I on the basis of the Sturm-Liouville theory, we note that our *two*-dimensional system (10.4) is not a Sturm-Liouville system. Nevertheless, like the Sturm-Liouville system it is *self-adjoint* and, under very general conditions, the eigenfunctions of self-adjoint operators are complete.[27]

Actually, expansion of a given function, say $f(\xi, \eta)$, in terms of the eigenfunctions (10.13) can be justified *without* alluding to these more general theorems. Specifically, the "separated" systems (10.7) *are* Sturm-Liouville systems. Recall (5.8) and (5.9) of PART I: in (10.7a), "p" = "r" = 1, "q" = 0 and "λ" = k^2; in (10.7b), "p" = "r" = 1, "q" = $-k^2$ and "λ" = λ; also, the boundary conditions of (10.7) are of suitable form. Thus, regarding η as fixed, we may expand a given function $f(\xi, \eta)$ in terms of the $\sin(m\pi\xi/\mathrm{a})$ eigenfunctions of (10.7a). The coefficients of this expansion will be functions of η, which may then be expanded in terms of the $\sin(n\pi\eta/\mathrm{b})$ eigenfunctions of (10.7b). The end result will be identical to that obtained by direct expansion of f in terms of the eigenfunctions (10.13) of the two-dimensional system (10.4).

EXERCISES

10.1. At first glance, (10.15) may seem unjustifiable since only "reasonably well behaved" functions may be expanded in terms of the ϕ_{mn} eigenfunctions. As discussed in Section 3 of PART I, however, it is to be understood *symbolically*, in the sense that

$$\lim_{M,N\to\infty} \int_0^{\mathrm{b}} \int_0^{\mathrm{a}} w_{M,N}(\xi - x, \eta - y) h(\xi, \eta)\, d\xi\, d\eta = h(x, y) \tag{A}$$

where

[27] See, for example, R. Courant and D. Hilbert, *Methods of Mathematical Physics*, Vol. I, Interscience Publishers, Inc., New York, 1953.

$$w_{M,N}(\xi - x, \eta - y) = \sum_{m=1}^{M} \sum_{n=1}^{N} \frac{4}{\text{ab}} \phi_{mn}(x, y)\phi_{mn}(\xi, \eta)$$

Verify that (A) is, in fact, satisfied.

10.2. Consider the boundary value problem $\nabla^2 u = \phi$ in the semicircular region $r < R, 0 < \theta < \pi$, with the boundary conditions $u(R, \theta) = f(\theta)$, $u(r, 0) = u(r, \pi) = 0$. The Green's function satisfies $\nabla^2 G = \delta$ in $\mathscr{S}$, with $G = 0$ on $\mathscr{C}$. The associated eigenvalue problem is

$$\nabla^2 u + \lambda u = 0 \text{ in } \mathscr{S}; \qquad u = 0 \text{ on } \mathscr{C} \tag{A}$$

(a) Show that the eigenfunctions and eigenvalues of (A) are

$$\phi_{mn}(r, \theta) = \sin m\theta \, J_m\left(\frac{z_{mn} r}{R}\right)$$

$$\lambda_{mn} = \left(\frac{z_{mn}}{R}\right)^2$$

for $m, n = 1, 2, 3, \ldots$, where J_m is the Bessel function of first kind and order m, and $z_{m1}, z_{m2}, \ldots$ are the positive zeros of J_m, i.e., $J_m(z_{mn}) = 0$ for $n = 1, 2, \ldots$. *Hint:* The general solution of $x^2 y'' + xy' + (k^2 x^2 - n^2)y = 0$ is $y = AJ_n(kx) + BY_n(kx)$, where J_n and Y_n are the Bessel functions of first and second kind, and order n. The J_n's and Y_n's are bounded for all x, except at $x = 0$, where the Y_n's $\to -\infty$.

(b) Show that the above eigenfunctions are mutually orthogonal; i.e., $(\phi_{jk}, \phi_{mn}) = 0$ if $j \neq m$ and/or $k \neq n$. *Hint:*

$$\int_0^R r J_m\left(\frac{z_{mj} r}{R}\right) J_m\left(\frac{z_{mk} r}{R}\right) dr = 0 \qquad \text{for } j \neq k$$

(c) Deduce that

$$G(\rho, \vartheta; r, \theta) = -\sum_m \sum_n \frac{\sin m\vartheta \sin m\theta \, J_m(z_{mn}\rho/R) J_m(z_{mn} r/R)}{\pi z_{mn}^2 [J_m'(z_{mn})]^2/4}$$

and thus write down the solution of the original boundary value problem. *Hint:*

$$\int_0^R r\left[J_m\left(\frac{z_{mn} r}{R}\right)\right]^2 dr = \frac{R^2 [J_m'(z_{mn})]^2}{2}$$

[A very clear discussion on the application of Fourier-Bessel series to boundary value problems can be found in Chapter 8 of R. V. Churchill's *Fourier Series and Boundary Value Problems*, 2nd ed., McGraw-Hill Book Company, New York, 1963.]

10.3. Extend the ideas of Example 1 to the case where the region is *three-dimensional*, namely $0 < x < \mathrm{a}, 0 < y < \mathrm{b}, 0 < z < \mathrm{c}$. Compute the Green's function.

11. ADDITIONAL EXAMPLES

In the preceding ten sections we restricted our discussion to the case of second order equations in two independent variables. The purpose of this section is to show, by means of a few examples, that extension of the method to higher dimensions, and higher order equations, is entirely straightforward.

Example 1. *Laplace Operator in Three Dimensions.* We consider the boundary value problem

$$\nabla^2 u = \Phi \text{ in } \mathscr{V}; \qquad u = f \text{ on } \mathscr{S} \tag{11.1}$$

where $\mathscr{V}$ is the interior of a sphere of radius R, and $\mathscr{S}$ is its surface.

As usual, we start by integrating the product $G\mathbf{L}u$ by parts, over the region of interest:

$$\iiint_{\mathscr{V}} G\mathbf{L}u \, d\tau = \text{boundary terms} + \iiint_{\mathscr{V}} u\mathbf{L}^* G \, d\tau \tag{11.2}$$

where $d\tau$ is a differential element of volume, and $d\sigma$ will be used to denote a differential element of surface area. In this case it is unnecessary to work out the details of the integration by parts, since we recognize that the required formula (11.2) is none other than Green's theorem,

$$\iiint_{\mathscr{V}} G\nabla^2 u \, d\tau = \iint_{\mathscr{S}} (Gu_n - uG_n) \, d\sigma + \iiint_{\mathscr{V}} u\nabla^2 G \, d\tau \tag{11.3}$$

Setting

$$\mathbf{L}^* G = \nabla^2 G = \delta(\xi - x, \eta - y, \zeta - z) \text{ in } \mathscr{V} \tag{11.4a}$$

$$G = 0 \text{ on } \mathscr{S} \tag{11.4b}$$

and noting that $\nabla^2 u = \Phi$, (11.3) becomes

$$u(x, y, z) = \iint_{\mathscr{S}} fG_n \, d\sigma + \iiint_{\mathscr{V}} G\Phi \, d\tau \tag{11.5}$$

where it now remains to compute G, as prescribed by (11.4).

As before, it is convenient to seek G in the form

$$G(\xi, \eta, \zeta; x, y, z) = U(\xi, \eta, \zeta; x, y, z) + g(\xi, \eta, \zeta; x, y, z) \tag{11.6}$$

where the principal solution U satisfies

$$\mathbf{L}^* U = \nabla^2 U = \delta(\xi - x, \eta - y, \zeta - z) \tag{11.7}$$

subject to no particular boundary conditions, and the regular part g is to satisfy the homogeneous equation

$$\mathbf{L}^* g = \nabla^2 g = 0 \tag{11.8}$$

subject to boundary conditions which are such that $G = U + g = 0$ on $\mathscr{S}$, as required by (11.4b).

First consider U. Expressing the Laplacian in terms of *spherical polar coordinates* r, ϕ, θ with their origin at the singular point x, y, z we have

$$\nabla^2 U = \frac{1}{r^2}\frac{\partial}{\partial r}\left(r^2 \frac{\partial U}{\partial r}\right) + \frac{1}{r^2 \sin\phi}\frac{\partial}{\partial \phi}\left(\sin\phi \frac{\partial U}{\partial \phi}\right) + \frac{1}{r^2 \sin^2\phi}\frac{\partial^2 U}{\partial \theta^2} \tag{11.9}$$

where

$$r = \sqrt{(\xi - x)^2 + (\eta - y)^2 + (\zeta - z)^2} \tag{11.10}$$

Now, interpreting U, from (11.7), as the gravitational potential induced at ξ, η, ζ by a point unit mass located at x, y, z it is reasonable to seek U as a function of the radial distance r alone. Thus, for all $r > 0$ we have

$$\nabla^2 U = \frac{1}{r^2}\frac{d}{dr}\left(r^2 \frac{dU}{dr}\right) = 0 \tag{11.11}$$

Integrating,

$$U = \frac{A}{r} + B \tag{11.12}$$

Thus far we have stayed away from the singular point; i.e., (11.11) is valid for $r > 0$. To determine the integration constants A and B, we now integrate (11.7) over an infinitesimal sphere of radius ϵ, centered at the singular point,

$$\iiint \nabla^2 U \, d\tau = \iiint \delta \, d\tau \tag{11.13}$$

Applying Gauss's divergence theorem to the left-hand integral, and noting that the right-hand side is unity, we have

$$\iint U_n \, d\sigma = 1 \tag{11.14}$$

But $U_n = dU/dr = -A/r^2$ is $-A/\epsilon^2$ on the surface of the ϵ-sphere. Thus,

$$-\frac{A}{\epsilon^2} 4\pi\epsilon^2 = 1$$

so that $A = -1/4\pi$, and B remains arbitrary. Setting $B = 0$ for convenience, we have

$$U = -\frac{1}{4\pi r} \tag{11.15}$$

in contrast with the result $U = (1/2\pi) \ln r$ obtained in the two-dimensional case. [A more direct derivation of (11.15) is outlined in Exercise 11.1.]

We now proceed with the calculation of G by the method of **images.** First, it will be convenient to change over from Cartesian coordinates to spherical polars (Fig. 11.1),

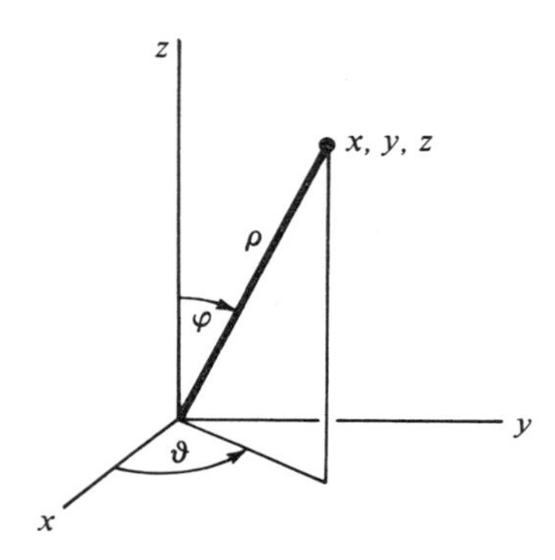

FIGURE 11.1 Spherical polars.

$$\begin{aligned} \xi &= \tilde{\rho} \cos \tilde{\vartheta} \sin \tilde{\varphi} & \qquad x &= \rho \cos \vartheta \sin \varphi \\ \eta &= \tilde{\rho} \sin \tilde{\vartheta} \sin \tilde{\varphi} & \qquad y &= \rho \sin \vartheta \sin \varphi \\ \zeta &= \tilde{\rho} \cos \tilde{\varphi} & \qquad z &= \rho \cos \varphi \end{aligned} \tag{11.16}$$

Now, recall that our starting point in the *two*-dimensional case was the basic geometrical statement that the circle $\mathscr{C}$ (Fig. 11.2), of radius R, is the locus of all points P such that

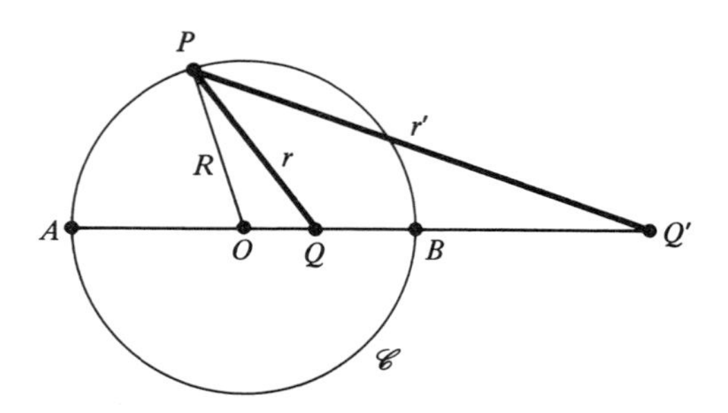

FIGURE 11.2 Image geometry.

$$\frac{QP}{Q'P} = \text{constant, say } \kappa \tag{11.17}$$

If we remove the restriction that P lie in the plane of the paper, then the result is a *sphere*. Recalling from (6.12) that $\kappa = OQ/R$, let us rewrite (11.17) as follows:

$$-\frac{1}{4\pi QP} + \frac{R/OQ}{4\pi Q'P} = 0 \text{ on } \mathscr{S} \tag{11.18}$$

Now let P be an arbitrary point *inside* the sphere. If we take Q to be our x, y, z field point, and P to be the variable ξ, η, ζ point, then QP is the distance r, and $OQ = \rho$. Defining $Q'P \equiv r'$, as well, we see that

$$-\frac{1}{4\pi QP} + \frac{R/OQ}{4\pi Q'P} = -\frac{1}{4\pi r} + \frac{R/\rho}{4\pi r'} \tag{11.19}$$

is equal to zero on $\mathscr{S}$. Furthermore, the first term is none other than U, and the second term is *harmonic* throughout the sphere; it is singular only at Q', which lies *outside* the sphere. But these are precisely the conditions required of G, so that we have

$$G = -\frac{1}{4\pi r} + \frac{R/\rho}{4\pi r'} \tag{11.20}$$

Interpreting this in terms of gravitational potential, we observe that the first term is the potential due to a point unit mass located at the ρ, φ, ϑ (x, y, z) field point, and the second term is the potential due to a "negative" point mass of strength R/ρ located at the *image point* $R^2/\rho, \varphi, \vartheta$.[28]

Finally, let us express r and r' in (11.20), in terms of our spherical polar coordinates (Exercise 11.2):

$$r = \sqrt{\tilde{\rho}^2 + \rho^2 - 2\tilde{\rho}\rho\cos\gamma} \tag{11.21a}$$

$$r' = \sqrt{\tilde{\rho}^2 + \left(\frac{R^2}{\rho}\right)^2 - 2\tilde{\rho}\frac{R^2}{\rho}\cos\gamma} \tag{11.21b}$$

where we have set

$$\cos\tilde{\varphi}\cos\varphi + \sin\tilde{\varphi}\sin\varphi\cos(\tilde{\vartheta} - \vartheta) \equiv \cos\gamma \tag{11.21c}$$

for brevity. With G given by (11.20) and (11.21), the solution is provided by (11.5), where we note that

$$\begin{aligned} G_n\big|_{\text{on } \mathscr{S}} &= G_{\tilde{\rho}}\big|_{\tilde{\rho}=R} \\ d\sigma &= (R\,d\tilde{\varphi})(R\sin\tilde{\varphi}\,d\tilde{\vartheta}) \\ d\tau &= (R\,d\tilde{\varphi})(R\sin\tilde{\varphi}\,d\tilde{\vartheta})(d\tilde{\rho}) \end{aligned}$$

[28] For a more general discussion of William Thomson's method of "reciprocal radii" in potential theory, we refer the reader to A. Sommerfeld's *Partial Differential Equations in Physics,* Academic Press, Inc., New York, 1949, pp. 139–142.

For the Laplace equation, where $\Phi = 0$, this yields the well-known **Poisson integral formula**

$$u(\rho, \varphi, \vartheta) = \frac{R(R^2 - \rho^2)}{4\pi} \int_0^{2\pi} \int_0^{\pi} \frac{f(\tilde{\varphi}, \tilde{\vartheta}) \sin \tilde{\varphi} \, d\tilde{\varphi} \, d\tilde{\vartheta}}{[R^2 + \rho^2 - 2R\rho \cos \gamma]^{3/2}} \tag{11.22}$$

Example 2. *Two- and Three-Dimensional Acoustics.* We will consider the acoustic equation

$$\mathbf{L}u = c^2 \nabla^2 u - u_{\tau\tau} = \phi \tag{11.23}$$

in both two and three space dimensions.

First, we compute principal solutions. For the **two-dimensional** case,

$$\mathbf{L}^* U = c^2(U_{\xi\xi} + U_{\eta\eta}) - U_{\tau\tau} = \delta(\eta - x, \eta - y, \tau - t) \tag{11.24}$$

To solve this, let us Laplace transform on the variable τ. That is, we multiply through by $\exp(-s\tau)\, d\tau$ and integrate from $\tau = 0$ to $\tau = \infty$, where s is the (complex) transform variable:

$$c^2 \int_0^\infty (U_{\xi\xi} + U_{\eta\eta}) e^{-s\tau} \, d\tau - \int_0^\infty U_{\tau\tau} e^{-s\tau} \, d\tau = \int_0^\infty \delta(\xi - x, \eta - y, \tau - t) e^{-s\tau} \, d\tau \tag{11.25}$$

This reduces to

$$c^2(\bar{U}_{\xi\xi} + \bar{U}_{\eta\eta}) - s^2 \bar{U} = e^{-st} \delta(\xi - x, \eta - y) \tag{11.26}$$

where

$$\bar{U} = \int_0^\infty U(\xi, \eta, \tau\,; x, y, t) e^{-s\tau} \, d\tau \tag{11.27}$$

is the *Laplace transform* of U, and where we have taken both U and U_τ to be zero at $\tau = 0$ in the evaluation of the boundary terms arising from repeated integration by parts of the middle integral in (11.25). We defend these initial values as being reasonable on the grounds that the delta function input does not act until $\tau = t$, so that $U \equiv 0$ for all $\tau < t$. (Recall that the boundary and initial conditions imposed in the calculation of principal solutions are quite arbitrary anyway, and are generally selected based upon physical motivation or mathematical convenience.) If we require, further, that U (and hence $\bar{U}$) be symmetric about the x, y singular point, then (11.26) reduces to the *ordinary* differential equation

$$c^2\left(\bar{U}_{rr} + \frac{1}{r} \bar{U}_r\right) - s^2 \bar{U} = e^{-st} \delta(\xi - x, \eta - y) \tag{11.28}$$

where r is the radial distance from x, y:

$$r = \sqrt{(\xi - x)^2 + (\eta - y)^2} \tag{11.29}$$

For $r > 0$, we have from (11.28)

$$r\bar{U}_{rr} + \bar{U}_r - \left(\frac{s}{c}\right)^2 r\bar{U} = 0 \tag{11.30}$$

which is identical, in form, to equation (5.13)—except for the minus sign. Linearly independent solutions of (11.30) are given by $I_0[(s/c)r]$ and $K_0[(s/c)r]$, the *modified zeroth order Bessel functions, of the first and second kind* respectively. They admit expansions of the form[29]

$$I_0(x) = 1 + \left(\frac{x}{2}\right)^2 + \frac{(x/2)^4}{(2!)^2} + \frac{(x/2)^6}{(3!)^2} + \cdots$$
$$= \sum_{n=0}^{\infty} \frac{(x/2)^{2n}}{(n!)^2} \sim 1 \text{ as } x \to 0 \tag{11.31a}$$

$$K_0(x) = -\left(\gamma + \ln\frac{x}{2}\right)I_0(x) + \sum_{n=1}^{\infty}\frac{(x/2)^{2n}}{(n!)^2}\left\{1 + \frac{1}{2} + \frac{1}{3} + \cdots + \frac{1}{n}\right\}$$
$$\sim -\ln x \text{ as } x \to 0 \tag{11.31b}$$

where γ is *Euler's constant* 0.5772157. . . .

Of these two, we discard I_0 since it is regular at $r = 0$, and tentatively set

$$\bar{U} = AK_0\left(\frac{sr}{c}\right) \tag{11.32}$$

To compute A, we integrate (11.26) over a disk of arbitrarily small radius ϵ, centered at x, y. Applying the Gauss divergence theorem to the integral of $c^2\nabla^2\bar{U}$ (recall Footnote 9), and noting that the integral of δ is unity, we obtain

$$c^2\{\bar{U}_r|_{r=\epsilon}\}2\pi\epsilon - s^2\iint \bar{U}\,d\sigma = e^{-st} \tag{11.33}$$

According to (11.31) and (11.32), however,

$$\bar{U} \sim -A\ln r$$
$$\bar{U}_r \sim -\frac{A}{r}$$

so (11.33) becomes

$$c^2\left(-\frac{A}{\epsilon}\right)2\pi\epsilon + O(\epsilon^2\ln\epsilon) = e^{-st} \tag{11.34}$$

and, letting $\epsilon \to 0$,

$$A = -\frac{e^{-st}}{2\pi c^2} \tag{11.35}$$

[29] McLachlan's *Bessel Functions for Engineers,* pp. 200 and 203.

Thus,

$$\bar{U} = -\frac{e^{-st}K_0(sr/c)}{2\pi c^2} \tag{11.36}$$

From Laplace transform tables,[30] the inverse is given by

$$U = \begin{cases} 0, & r > c(\tau - t) \quad (11.37a) \\ -\dfrac{1}{2\pi c\sqrt{c^2(\tau - t)^2 - r^2}}, & r < c(\tau - t) \quad (11.37b) \end{cases}$$

as sketched in Fig. 11.3.

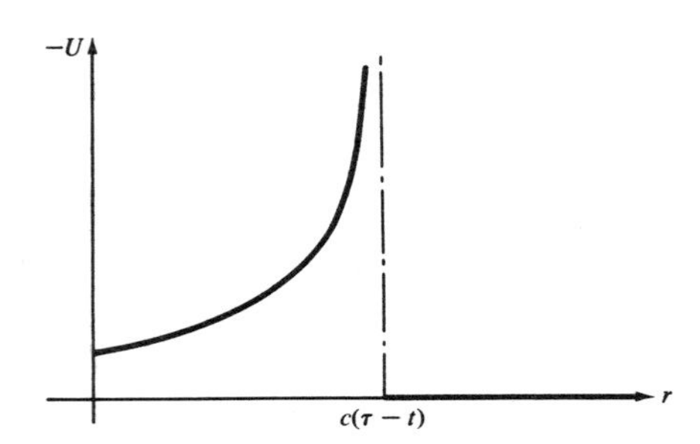

FIGURE 11.3 Principal solution of two-dimensional wave equation; Eq. (11.37).

For the **three-dimensional** case, a similar calculation (Exercise 11.7) leads to

$$U = -\frac{\delta[\tau - t - (r/c)]}{4\pi c^2 r} \tag{11.38}$$

where

$$r = \sqrt{(\xi - x)^2 + (\eta - y)^2 + (\zeta - z)^2} \tag{11.39}$$

Comparative interpretation of the principal solutions (11.37) and (11.38) is extremely interesting. The solution (11.38), for example, represents the disturbance induced by a "firecracker of unit strength" which is set off at $r = 0$ (i.e., at the field point x, y, z) and $\tau = t$ in unbounded space. We see that it is confined to an infinitely thin spherical shell, the radius of which increases steadily with the speed of sound c, and its strength decreases like $1/r$ as r increases. Thus, an observer at a distance r from the source hears nothing until

[30] For example, Churchill's *Operational Mathematics*, 2nd ed., p. 330, together with the Translation Theorem on p. 23.

$\tau - t = r/c$; at that instant he hears a "bang," which is followed by silence. In the two-dimensional case, by contrast, the "bang" at $\tau - t = r/c$ does not die out instantly; instead, it decays according to (11.37b).[31] As Mackie points out,[32] "*. . . we might reflect with some gratitude that our existence is in a three-dimensional world! If it were in two dimensions, any form of aural communication would be extremely tedious as individual sounds cound hardly be distinguished, each one dying away only gradually and so becoming mixed up with every other.*"

The reason for this profound difference between the two- and three-dimensional cases is easily explained. Consider the two-dimensional disturbance to be caused by a straight infinite wire which is "exploded" at time $\tau = t$. The disturbance at a point P which is at a perpendicular distance r from the wire is described, as a function of time, by (11.37). Actually, the wire consists of a linear *distribution of point sources*. At $\tau = t$, each source sends out a disturbance which expands spherically, according to (11.38). The first disturbances to reach P are those emitted from the neighborhood of point Q (Fig. 11.4). Subsequent disturbances continue to arrive for all $\tau > t$. They become weaker and weaker since they were emitted at points on the wire which are farther and farther away from P. This is in accordance with (11.37), as sketched in Fig. 11.3.

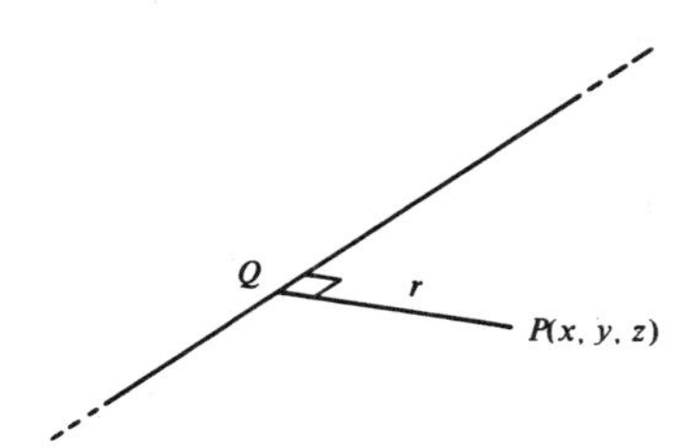

FIGURE 11.4 Exploding wire.

[31] The two-dimensional wave equation also governs the propagation of surface waves in shallow water of uniform depth. If we drop a stone into a shallow pond, we observe that the leading ripple spreads out in a circle; its radius increases at a steady rate, and its height decreases as $1/r$. Furthermore, the water enclosed by this ripple remains in motion even after its passage, although this motion decays with time.

[32] A. G. Mackie, *Boundary Value Problems*, Hafner Publishing Company, New York, 1965, p. 176. See also the earlier comments of R. Courant, *Modern Mathematics for the Engineer*, Vol. 1, edited by E. F. Beckenbach, McGraw-Hill Book Company, New York, 1956, p. 100.

Mathematically, what we are saying is that the two-dimensional result (11.37) can be obtained from the three-dimensional one by superposition:

$$U(\xi, \eta, \tau; x, y, t) = -\int_{-\infty}^{\infty} \frac{\delta[\tau - t - (1/c)\sqrt{(\xi - x)^2 + (\eta - y)^2 + \zeta^2}]}{4\pi c^2\sqrt{(\xi - x)^2 + (\eta - y)^2 + \zeta^2}}\, d\zeta \tag{11.40}$$

Noting that the integral is twice that from 0 to ∞, and setting

$$(\xi - x)^2 + (\eta - y)^2 + \zeta^2 \equiv c^2\mu^2$$

we obtain

$$U(\xi, \eta, \tau; x, y, t) = -\frac{1}{2\pi c}\int_{r/c}^{\infty} \frac{\delta[\mu - (\tau - t)]}{\sqrt{c^2\mu^2 - r^2}}\, d\mu \tag{11.41}$$

where

$$r = \sqrt{(\xi - x)^2 + (\eta - y)^2}$$

Now, if $\tau - t < r/c$ then the action of the delta function is outside the interval of integration, and the integral is zero. This agrees with (11.37a). If, on the other hand, $\tau - t > r/c$ then we obtain

$$-\frac{1}{2\pi c\sqrt{c^2(\tau - t)^2 - r^2}}$$

again in agreement with (11.37b).

Similarly, we can obtain the *one*-dimensional result (5.25) by the superposition of two-dimensional results (Exercise 11.8).

Now, as in the case of the one-dimensional wave equation (end of Section 5), we observe that additional principal solutions may be obtained from (11.37) and (11.38) if we replace $\tau - t$ by $t - \tau$. This produces the two- and three-dimensional **backward running principal solutions**

$$U = \begin{cases} 0, & r > c(t - \tau) \quad (11.42a) \\ -\dfrac{1}{2\pi c\sqrt{c^2(t - \tau)^2 - r^2}}, & r < c(t - \tau) \quad (11.42b) \end{cases}$$

and

$$U = -\frac{\delta[t - \tau - (r/c)]}{4\pi c^2 r} \tag{11.43}$$

respectively.

Application. Let us apply these results to the two-dimensional acoustics problem,

$$\mathbf{L}u = c^2\nabla^2 u - u_{\tau\tau} = 0 \tag{11.44a}$$

over $-\infty < \xi < \infty$, $-\infty < \eta < \infty$, $0 < \tau < T$ (where T exceeds any time t at which the solution may be sought), subject to the initial conditions

$$u(\xi, \eta, 0) = f(\xi, \eta) \tag{11.44b}$$

$$u_\tau(\xi, \eta, 0) = h(\xi, \eta) \tag{11.44c}$$

Integrating $G\mathbf{L}u$ by parts,

$$\int_0^T \int_{-\infty}^{\infty} \int_{-\infty}^{\infty} G\mathbf{L}u \, d\xi \, d\eta \, d\tau = \int_{-\infty}^{\infty} \int_{-\infty}^{\infty} (uG_\tau - Gu_\tau)\Big|_{\tau=0}^{\tau=T} d\xi \, d\eta + \int_0^T \int_{-\infty}^{\infty} \int_{-\infty}^{\infty} u\mathbf{L}^* G \, d\xi \, d\eta \, d\tau \tag{11.45}$$

Since u and u_τ are not specified at $\tau = T$, we require that $G = G_\tau = 0$ at $\tau = T$. But, since T is an *arbitrary* time, greater than t, it follows that G is identically zero for *all* $\tau > t$. Thus, the boundary value problem on the Green's function is

$$\mathbf{L}^* G = c^2 \nabla^2 G - G_{\tau\tau} = \delta(\xi - x, \eta - y, \tau - t) \tag{11.46a}$$

over $-\infty < \xi < \infty$, $-\infty < \eta < \infty$, $0 < \tau < T$, subject to the condition that

$$G(\xi, \eta, \tau; x, y, t) = 0 \text{ for all } \tau > t \tag{11.46b}$$

Using (11.44) and (11.46), the solution is given by (11.45), in the form

$$u(x, y, t) = \int_{-\infty}^{\infty} \int_{-\infty}^{\infty} [f(\xi, \eta) G_\tau(\xi, \eta, 0; x, y, t) - h(\xi, \eta) G(\xi, \eta, 0; x, y, t)] \, d\xi \, d\eta \tag{11.47}$$

Inspecting (11.46), we observe that the Green's function is simply the backward running principal solution (11.42). For the case where $f(\xi, \eta) \equiv 0$, (11.47) reduces to

$$u(x, y, t) = \frac{1}{2\pi c} \iint_{\mathscr{S}(t)} \frac{h(\xi, \eta) \, d\xi \, d\eta}{\sqrt{c^2 t^2 - r^2}} \tag{11.48}$$

where the region of integration $\mathscr{S}(t)$ is the disk $r < ct$, centered at x, y. That is, the disturbance at x, y, t is due only to h values from within the domain of dependence $r < ct$.

Rewriting the integral in (11.48) in terms of polar coordinates r, θ

$$\frac{1}{2\pi c} \int_0^{2\pi} \int_0^{ct} \frac{hr \, dr \, d\theta}{\sqrt{(ct - r)(ct + r)}}$$

we note that the integrand has a square-root singularity at the limit $r = ct$, i.e., the integrand is $O(ct - r)^{-1/2}$ there. This is *weaker* than first order, i.e., $O(ct - r)^{-1}$, so that the integral does exist. On the other hand, the G_τ term in (11.47) is $O(ct - r)^{-3/2}$, which is *stronger* than first order, and hence *not* integrable. Thus, (11.47) is not valid in the event that f is nonzero. The source of the difficulty is not very mysterious; the fact is that the integration by parts which produced the G_τ term is not permissible,[33] and special care is required; see, for example, Mackie's *Boundary Value Problems*, p. 192.

Application. Next, consider the problem

$$c^2\nabla^2 u - u_{\tau\tau} = \phi(\xi, \eta, \zeta, \tau) \tag{11.49}$$

in unbounded *three*-dimensional space, where $u = u_\tau = 0$ at $\tau = 0$. We find easily (Exercise 11.10) that

$$u(x, y, z, t) = -\iiiint \frac{\delta[t - \tau - (r/c)]}{4\pi c^2 r} \phi(\xi, \eta, \zeta, \tau)\, d\xi\, d\eta\, d\zeta\, d\tau \tag{11.50}$$

$$= -\iiint \frac{\phi[\xi, \eta, \zeta, t - (r/c)]}{4\pi c^2 r}\, d\xi\, d\eta\, d\zeta \tag{11.51}$$

where the limits of integration are $-\infty$ to ∞ on the space variables, and 0 to ∞ on τ. Observe that the time argument of ϕ is *retarded* by an amount r/c since a ϕ disturbance at ξ, η, ζ takes that much time to reach x, y, z. The result (11.50) is therefore referred to as a **retarded potential.**

For definiteness, suppose

$$\phi = \delta[\xi - a(\tau), \eta, \zeta] \tag{11.52}$$

Physically, this corresponds to a unit source which moves along the ξ axis according to $\xi = a(\tau)$. From (11.50),

$$u(x, y, z, t) = -\iiiint \frac{\delta[t - \tau - (r/c)]}{4\pi c^2 r} \delta[\xi - a(\tau), \eta, \zeta]\, d\xi\, d\eta\, d\zeta\, d\tau \tag{11.53}$$

Now, we should be able to evaluate this quadruple integral explicitly, since the integrand contains, in effect, a four-dimensional delta function (i.e., the product of one- and three-dimensional delta functions). However, the evaluation is not trivial because of the complicated arguments, $t - \tau - (r/c)$ and $\xi - a(\tau)$, in the delta functions. Following the discussion in Mathews and Walker,[34] we simplify the arguments by the change of variables

[33] To illustrate this point with a simple example, we point out that although the integral $\int_0^1 e^x x^{-1/2}\, dx$ does exist, the integration by parts

$$\int_0^1 e^x x^{-1/2}\, dx = (e^x x^{-1/2})\Big|_0^1 + \int_0^1 e^x x^{-3/2}\, dx$$

is *not* valid since the terms on the right do not exist.

[34] J. Mathews and R. L. Walker, *Mathematical Methods of Physics*, W. A. Benjamin, Inc., New York, 1965, pp. 267–268.

$$
\begin{aligned}
t - \tau - \frac{r}{c} &= v_1 \\
\xi - a(\tau) &= v_2 \\
\eta &= v_3 \\
\zeta &= v_4
\end{aligned}
\tag{11.54}
$$

Now, we recall from the integral calculus that under a change of variables the hypervolume element $d\xi\, d\eta\, d\zeta\, d\tau$ becomes

$$
d\xi\, d\eta\, d\zeta\, d\tau = \left|\frac{\partial(\xi\eta\zeta\tau)}{\partial(v_1 v_2 v_3 v_4)}\right| dv_1\, dv_2\, dv_3\, dv_4 \tag{11.55}
$$

where

$$
\frac{\partial(\xi\eta\zeta\tau)}{\partial(v_1 v_2 v_3 v_4)} = \begin{vmatrix}
\frac{\partial\xi}{\partial v_1} & \frac{\partial\xi}{\partial v_2} & \frac{\partial\xi}{\partial v_3} & \frac{\partial\xi}{\partial v_4} \\
\frac{\partial\eta}{\partial v_1} & \frac{\partial\eta}{\partial v_2} & \frac{\partial\eta}{\partial v_3} & \frac{\partial\eta}{\partial v_4} \\
\frac{\partial\zeta}{\partial v_1} & \frac{\partial\zeta}{\partial v_2} & \frac{\partial\zeta}{\partial v_3} & \frac{\partial\zeta}{\partial v_4} \\
\frac{\partial\tau}{\partial v_1} & \frac{\partial\tau}{\partial v_2} & \frac{\partial\tau}{\partial v_3} & \frac{\partial\tau}{\partial v_4}
\end{vmatrix} \tag{11.56}
$$

is the *Jacobian determinant* of ξ, η, ζ, τ with respect to v_1, v_2, v_3, v_4 and the vertical bars in (11.55) denote absolute value. Thus, (11.53) becomes

$$
\begin{aligned}
u(x, y, z, t) &= -\iiiint \frac{\delta(v_1)\delta(v_2, v_3, v_4)}{4\pi c^2 r}\left|\frac{\partial(\xi\eta\zeta\tau)}{\partial(v_1 v_2 v_3 v_4)}\right| dv_1\, dv_2\, dv_3\, dv_4 \\
&= -\left\{\frac{1}{4\pi c^2 r}\left|\frac{\partial(\xi\eta\zeta\tau)}{\partial(v_1 v_2 v_3 v_4)}\right|\right\}\Bigg|_{v_1=v_2=v_3=v_4=0}
\end{aligned}
\tag{11.57}
$$

Actually, it is convenient to use the known property of Jacobians,

$$
\frac{\partial(\xi\eta\zeta\tau)}{\partial(v_1 v_2 v_3 v_4)} = \left[\frac{\partial(v_1 v_2 v_3 v_4)}{\partial(\xi\eta\zeta\tau)}\right]^{-1} \tag{11.58}
$$

since, for example, $\partial v_2/\partial\tau$ is more easily computed than $\partial\tau/\partial v_2$, because of the form of (11.54). Straightforward calculation yields

$$
\frac{\partial(v_1 v_2 v_3 v_4)}{\partial(\xi\eta\zeta\tau)} = \frac{a'(\tau)[a(\tau) - x]}{cr} + 1
$$

so that

$$
u(x, y, z, t) = -\frac{1}{4\pi c}\frac{1}{|a'(\tau)[a(\tau) - x] + cr|} \tag{11.59a}
$$

where

$$r = \sqrt{[a(\tau) - x]^2 + y^2 + z^2} \tag{11.59b}$$

and τ is defined *implicitly*, in terms of x, y, z, t, by

$$t - \tau - \frac{r}{c} = 0 \tag{11.59c}$$

COMMENT 1. In electrodynamics, (11.59a) is known as the **Lienard-Wiechert potential.** In that case c is the speed of light, and the absolute value signs can be omitted since $|a(\tau) - x| \leq r$ and $|a'(\tau)|$ must be $< c$.

COMMENT 2. At first glance, some of our results may appear to disagree with corresponding results in other books. For example, the (forward running) principal solution of the three-dimensional wave equation is given by Mathews and Walker in the form

$$-\frac{c}{4\pi r}\delta[r - c(\tau - t)] \equiv \text{“}U_{M\&W}\text{”}$$

which is apparently not the same as our result (11.38). However, first we note that we have used $\mathbf{L}^* U = (c^2\nabla^2 - \partial^2/\partial\tau^2)U = \delta$, whereas Mathews and Walker use $[\nabla^2 - (1/c^2)\partial^2/\partial\tau^2]U_{M\&W} = \delta$,[35] so that the quantity

$$\frac{1}{c^2}U_{M\&W} = -\frac{\delta[r - c(\tau - t)]}{4\pi c r}$$

should agree with (11.38). Well, we see that it *does*, if we note (Exercise 3.9 of PART I) that

$$\begin{aligned}\delta[r - c(\tau - t)] &= \delta[c(\tau - t) - r] \\ &= \frac{1}{c}\delta\left(\tau - t - \frac{r}{c}\right)\end{aligned}$$

This illustrates the fact that *care must be used in regard to the arguments of delta functions.*

COMMENT 3. Finally, observe that in changing over to the v variables, according to (11.54), we did not bother spelling out the new limits of integration. They are, in fact, immaterial—except insofar as whether or not the domain of integration contains the point $v_1 = v_2 = v_3 = v_4 = 0$, at which the delta functions act. [If it does *not*, then the right side of (11.59a) should be replaced by *zero*.] The question is as follows: Given the quantities x, y, z, t, is there a (real) solution ξ, η, ζ, τ of the equations

[35] Furthermore, whereas we use δ on the right-hand side, some authors prefer $-4\pi\delta$.

$$v_1 = t - \tau - \frac{r}{c} = 0 \tag{11.60a}$$

$$v_2 = \xi - a(\tau) = 0 \tag{11.60b}$$

$$v_3 = \eta = 0 \tag{11.60c}$$

$$v_4 = \zeta = 0 \tag{11.60d}$$

where $-\infty < \xi < \infty, -\infty < \eta < \infty, -\infty < \zeta < \infty, 0 < \tau < \infty$? Well, (11.60b–d) yield $\eta = \zeta = 0$ and $\xi = a(\tau)$. Inserting these into (11.60a), the question reduces to the existence of a positive solution τ of the equation

$$c(t - \tau) = \sqrt{[a(\tau) - x]^2 + y^2 + z^2} \tag{11.61}$$

If, for example, $a(\tau) =$ constant $= 0$ then, clearly, (11.61) will *not* have a positive solution τ if x, y, z, t are such that

$$ct < \sqrt{x^2 + y^2 + z^2} \tag{11.62}$$

and it would follow that $u(x, y, z, t) = 0$. If, on the other hand, (11.61) does admit a positive solution τ, then $u(x, y, z, t)$ is given by (11.59a).

Example 3. *Biharmonic Equation.* As our final example, we will consider the inhomogeneous **biharmonic** equation

$$\mathbf{L}u = \nabla^4 u = \phi \tag{11.63}$$

over a two-dimensional region $\mathscr{S}$, together with suitable boundary conditions on $\mathscr{C}$. By ∇^4 we mean $\nabla^2\nabla^2$, where ∇^2 is the two-dimensional Laplacian.

In physical terms, (11.63) governs the *lateral deflection of an elastic plate*, subjected to the load distribution ϕ[36] over $\mathscr{S}$, and the prescribed boundary conditions on $\mathscr{C}$.

First, we integrate $G\mathbf{L}u$ by parts. Whereas we could "start from scratch," proceeding as in equations (2.2), it is easier to apply Green's theorem twice:

$$\begin{aligned}\iint_{\mathscr{S}} G\mathbf{L}u\, d\sigma &= \iint_{\mathscr{S}} G\nabla^2(\nabla^2 u)\, d\sigma \\ &= \int_{\mathscr{C}} [G(\nabla^2 u)_n - (\nabla^2 u)G_n]\, ds + \iint_{\mathscr{S}} \nabla^2 u \nabla^2 G\, d\sigma \\ &= \int_{\mathscr{C}} [G(\nabla^2 u)_n - (\nabla^2 u)G_n + (\nabla^2 G)u_n - u(\nabla^2 G)_n]\, ds \\ &\quad + \iint_{\mathscr{S}} u\nabla^4 G\, d\sigma \end{aligned} \tag{11.64}$$

[36] More precisely, ϕ is the applied normal force per unit area, divided by a physical constant which involves the thickness and elastic constants of the plate.

If, for example, the plate is clamped along its edge, such that $u = u_n = 0$ on $\mathscr{C}$, then (11.64) yields

$$u(x, y) = \iint_{\mathscr{S}} G(\xi, \eta; x, y)\phi(\xi, \eta)\, d\sigma \tag{11.65}$$

where the Green's function satisfies

$$\mathbf{L}^* G = \nabla^4 G = \delta(\xi - x, \eta - y) \text{ in } \mathscr{S} \tag{11.66a}$$

$$G = G_n = 0 \text{ on } \mathscr{C} \tag{11.66b}$$

Comparing (11.66) with (11.63), it is clear that G is the deflection at ξ, η due to a unit load at x, y. Observing that the operator $\mathscr{G}$ is self-adjoint, it follows that G is symmetric:

$$G(\xi, \eta; x, y) = G(x, y; \xi, \eta) \tag{11.67}$$

Thus G in (11.65) can also be interpreted as the deflection at x, y due to a unit load at ξ, η, so that the integral represents the superposition of the deflections caused by all the incremental loads $\phi(\xi, \eta)\, d\sigma$.

As for the calculation of G, we express

$$G(\xi, \eta; x, y) = U(\xi, \eta; x, y) + g(\xi, \eta; x, y) \tag{11.68}$$

where the principal solution U satisfies

$$\mathbf{L}^* U = \nabla^4 U = \delta(\xi - x, \eta - y) \tag{11.69}$$

over $\mathscr{S}$, subject to no particular boundary conditions, and the regular part g satisfies

$$\mathbf{L}^* g = \nabla^4 g = 0 \tag{11.70a}$$

over $\mathscr{S}$, subject to the boundary conditions

$$g = -U \tag{11.70b}$$

$$g_n = -U_n \tag{11.70c}$$

on $\mathscr{C}$, so that (11.66b) is satisfied.

To compute U, we express ∇^4 in terms of polar coordinates r, θ at the singular point x, y. Noting that there are no particular boundary conditions on U, it is convenient to require it to be symmetric about x, y, i.e., independent of θ. For $r > 0$, then,

$$\begin{aligned} \nabla^4 U &= \nabla^2 \nabla^2 U \\ &= \frac{1}{r}\frac{d}{dr}\left(r\frac{d}{dr}\right)\frac{1}{r}\frac{d}{dr}\left(r\frac{d}{dr}\right)U = 0 \end{aligned} \tag{11.71}$$

Integrating,

$$U = Ar^2 \ln r + Br^2 + C \ln r + D \tag{11.72}$$

To evaluate the constants, we integrate (11.69) over an arbitrarily small disk of radius ϵ, centered at x, y:

$$\iint \nabla^4 U \, d\sigma = \iint \delta \, d\sigma = 1 \tag{11.73}$$

The left-hand side is reduced by expressing $\nabla^4 = \nabla \cdot \nabla \nabla^2$ and applying the divergence theorem,

$$\text{LHS} = \iint \nabla \cdot (\nabla \nabla^2 U) d\sigma = \int \mathbf{n} \cdot (\nabla \nabla^2 U) \, ds$$

$$= \int (\mathbf{n} \cdot \nabla) \nabla^2 U \, ds = \int (\nabla^2 U)_n \, ds$$

But

$$(\nabla^2 U)_n = (\nabla^2 U)_r |_{r=\epsilon} = \frac{4A}{\epsilon}$$

so that

$$\frac{4A}{\epsilon}(2\pi\epsilon) = 1$$

and

$$A = \frac{1}{8\pi} \tag{11.74}$$

Since B, C, and D remain arbitrary, we take them to be zero, for convenience, and have

$$U = \frac{1}{8\pi} r^2 \ln r \tag{11.75}$$

Once the region $\mathscr{S}$ is specified, it remains to calculate the regular part g, according to (11.70). Depending upon the nature of $\mathscr{S}$, this may be accomplished by means of separation of variables, eigenfunction expansions, or integral transforms, for example. We merely note, in closing, that if $\mathscr{S}$ is the unit disk, the Green's function can be found in closed form,

$$G(\xi, \eta; x, y) = \frac{|\zeta - z|^2}{8\pi} \ln \left| \frac{\bar{\zeta} z - 1}{\zeta - z} \right| + \frac{1}{16\pi}(|\zeta|^2 - 1)(|z|^2 - 1) \tag{11.76}$$

where $\zeta = \xi + i\eta$, $\bar{\zeta} = \zeta - i\eta$ and $z = x + iy$.[37]

[37] Further discussion of the biharmonic equation may be found in S. Bergman and M. Schiffer, *Kernel Functions and Elliptic Differential Equations in Mathematical Physics,* Academic Press, Inc., New York, 1953. See also A. Kalnins, "On Fundamental Solutions and Green's Functions in the Theory of Elastic Plates," *J. Applied Mechanics,* March 1966, pp. 31–38, and the references therein.

EXERCISES

11.1. Derive the principal solution (11.15) more directly by starting with the law of gravitational attraction. (See Exercise 5.1.)

11.2. Derive equations (11.21) and (11.22).

11.3. What is the Green's function for the problem $u_{xx} + u_{yy} + u_{zz} = \phi$ in the half-space $x > 0$, with Dirichlet boundary conditions?

11.4. Verify, by means of a suitable limiting procedure, that the Green's function (11.20) for the Laplace operator in a sphere, with Dirichlet boundary conditions, does in fact reduce to the Green's function of Exercise 11.3 as the radius $R \to \infty$.

11.5. Obtain the principal solution $(1/2\pi) \ln r$ for the two-dimensional Laplace operator, by suitable integration of the three-dimensional result, $-1/4\pi r$.

11.6. Find the Green's function for the Laplace operator in a hemisphere, with Dirichlet boundary conditions, by means of a suitable image system.

11.7. Derive the (forward running) principal solution (11.38) of the three-dimensional wave equation. *Hint:* Starting with $c^2\nabla^2 U - U_{\tau\tau} = \delta(\xi - x, \eta - y, \zeta - z, \tau - t)$, Laplace transform on τ, and obtain $(r\bar{U})_{,rr} - (s/c)^2(r\bar{U}) = (r/c^2)e^{-st}\delta(\xi - x, \eta - y, \zeta - z)$. Thus show that $\bar{U} = -e^{-[t+(r/c)]s}/4\pi c^2 r$, and that the inverse transform yields (11.38).

11.8. Rework Exercise 11.7 using a *four-dimensional Fourier transform* instead. [Note the transform and inversion formulas

$$\hat{f}(\xi', \eta', \zeta', \tau') = \iiiint f(\xi, \eta, \zeta, \tau)e^{-i(\xi\xi' + \eta\eta' + \zeta\zeta' + \tau\tau')}\, d\xi\, d\eta\, d\zeta\, d\tau$$

$$f(\xi, \eta, \zeta, \tau) = \frac{1}{(2\pi)^4}\iiiint \hat{f}(\xi', \eta', \zeta', \tau')e^{i(\xi\xi' + \eta\eta' + \zeta\zeta' + \tau\tau')}\, d\xi'\, d\eta'\, d\zeta'\, d\tau'$$

respectively, where all integrals are from $-\infty$ to ∞.] First, show that

$$U(\xi, \eta, \zeta, \tau) = \frac{1}{(2\pi)^4}\iiint e^{i[(\xi - x)\xi' + (\eta - y)\eta' + (\zeta - z)\zeta']}\, d\xi'\, d\eta'\, d\zeta' \times \int \frac{e^{i(\tau - t)\tau'}}{(\tau')^2 - c^2[(\xi')^2 + (\eta')^2 + (\zeta')^2]}\, d\tau'$$

Changing the space variables from ξ', η', ζ' to spherical polars r', ϕ, θ [using the vector $(\xi - x, \eta - y, \zeta - z)$ as the polar axis, as shown in the figure] obtain

$$U(\xi, \eta, \zeta, \tau) = \frac{1}{(2\pi)^4} \int_0^{2\pi} \int_0^{\pi} \int_0^{\infty} e^{irr' \cos \phi} (r')^2 \sin \phi \, dr' \, d\phi \, d\theta$$

$$\times \int_{-\infty}^{\infty} \frac{e^{i(\tau - t)\tau'}}{(\tau')^2 - c^2 (r')^2} \, d\tau'$$

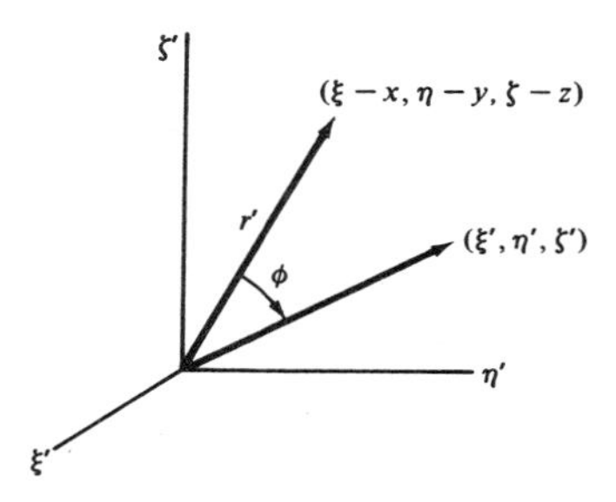

Evaluate the τ' integral by contour integration, completing the contour with a large semicircle in the upper half-plane for $\tau > t$, and in the lower half-plane for $\tau < t$. Noting the presence of two poles on the path of integration, show that the contour is to be deformed to pass *below* the two poles since we are after the *forward* running U. Completing the τ', ϕ, and θ integrations, obtain

$$U = \frac{1}{8\pi^2 cr} \int_0^{\infty} (e^{irr'} - e^{-irr'})(e^{i(\tau - t)cr'} - e^{-i(\tau - t)cr'}) \, dr'$$

$$= -\frac{1}{4\pi c^2 r} \left[\delta\left(\tau - t - \frac{r}{c}\right) - \delta\left(\tau - t + \frac{r}{c}\right) \right]$$

for $\tau > t$ and zero for $\tau < t$. (We have used formulas from the Exercises of Section 3, PART I.) Finally, show that the second delta function can be discarded, so that the result does agree with (11.38).

11.9. Obtain the principal solution (5.25) for the one-dimensional wave equation, by suitable integration of the two-dimensional result (11.37).

11.10. Provide the steps missing between (11.49) and (11.50).

11.11. Compute the solution of the system (11.44) for the case where $f \equiv 0$, where ξ is only *semi*-infinite, i.e., $0 < \xi < \infty$, where $u(0, \eta, \tau) = 0$.

11.12. Deduce (11.59a) from (11.53) as follows. Carrying out the integrations over the space variables, show that

$$u(x, y, z, t) = -\int_0^{\infty} \frac{\delta[\tau - t + (1/c)\sqrt{[a(\tau) - x]^2 + y^2 + z^2}]}{4\pi c^2 \sqrt{[a(\tau) - x]^2 + y^2 + z^2}} \, d\tau$$

and evaluate this with the help of Exercise 3.9 of PART I. (*Note that this line of approach is more direct* than the Jacobian method developed in the text, but the latter is more powerful; e.g., if the last two arguments of the three-dimensional delta function were considerably more complicated, we might very well *have* to resort to the Jacobian method.)

11.13. In the limit as $c \to \infty$, the wave equation becomes the Laplace (or Poisson) equation. Verify that the two- and three-dimensional principal solutions of the wave equation do reduce to the corresponding principal solutions of the Laplace equation in this limit.

11.14. Verify that (11.76) is, in fact, the desired Green's function, as claimed.

11.15. Compute the principal solution for the diffusion operator in *two* dimensions. Use this result to solve the following problem:

$$\mathbf{L}u = \kappa u_t - (u_{xx} + u_{yy}) = 0 \text{ in } y > 0, \ -\infty < x < \infty$$
$$u(x, y, 0) = 0 \text{ and } u(x, 0, t) = f(x, t) \text{ for } t > 0$$

11.16. The stream function Ψ of an axisymmetric, incompressible, irrotational flow is governed by the equation

$$\Psi_{rr} - \frac{1}{r}\Psi_r + \Psi_{xx} = 0$$

or

$$\mathbf{L}\left(\frac{\Psi}{r}\right) = \left(\nabla^2 - \frac{1}{r^2}\right)\frac{\Psi}{r} = 0$$

where x, r, θ are cylindrical coordinates, with x the axis of symmetry. Show that the principal solution U, satisfying $\mathbf{L}U = \delta(x - \xi, r - \rho)$, is

$$U(x, r; \xi, \rho) = -\frac{1}{2\pi}\sqrt{\frac{\rho}{r}}\,Q_{1/2}\left[1 + \frac{(x - \xi)^2 + (r - \rho)^2}{2r\rho}\right]$$

where $Q_{1/2}$ is a *Legendre function of second kind.* Physically, rU corresponds to the stream function at x, r due to a *circular vortex* of unit circulation and radius ρ, with its axis coinciding with the x axis, and located at $x = \xi$. *Hint:* Use a Fourier transform on x and a Hankel transform of order one (see Exercise 5.6) on r. The Fourier inversion can be carried out by the residue theorem and the Hankel inversion with the help of the formula

$$\int_0^\infty e^{-at}J_n(bt)J_n(ct)\,dt = \frac{1}{\pi\sqrt{bc}}\,Q_{n-(1/2)}\left[\frac{a^2 + b^2 + c^2}{2bc}\right]$$

12. SUMMARY

We have seen that the application of the Green's function method to partial differential equations is quite systematic, the procedure being essentially the same as for ordinary differential equations.

Given that

$$\mathbf{L}u = \phi \tag{12.1}$$

throughout an n-dimensional region $\mathscr{R}$, together with prescribed boundary conditions, we form the inner product[38]

$$(G, \mathbf{L}u) = \int \cdots \int_{\mathscr{R}} G\mathbf{L}u \, d\xi_1 \cdots d\xi_n \tag{12.2}$$

and integrate by parts:

$$(G, \mathbf{L}u) = \text{boundary terms} + (u, \mathbf{L}^* G) \tag{12.3}$$

Next, we require that G satisfy the partial differential equation

$$\mathbf{L}^* G = \delta(\xi_1 - x_1, \ldots, \xi_n - x_n) \tag{12.4}$$

throughout $\mathscr{R}$, together with homogeneous boundary conditions which eliminate those boundary terms in (12.3) which are unwelcome, i.e., those which contain combinations of u and its derivatives which are *not prescribed.*[39]

Using (12.1) and (12.4), the solution then follows from (12.3) in the form

$$u(x_1, \ldots, x_n) = -(\text{remaining boundary terms, if any}) + \int \cdots \int_{\mathscr{R}} G(\xi_1, \ldots, \xi_n; x_1, \ldots, x_n)\phi(\xi_1, \ldots, \xi_n) \, d\xi_1 \cdots d\xi_n \tag{12.5}$$

provided, of course, that we are able to find G.[40] Note that special care is required if $\mathscr{R}$ is unbounded.

To calculate the Green's function, it is generally convenient to split off the singular part. We express

$$G = U + g \tag{12.6}$$

where the singular part U satisfies

$$\mathbf{L}^* U = \delta(\xi_1, \ldots, \xi_n; x_1, \ldots, x_n) \tag{12.7}$$

subject to no particular boundary conditions, and the regular part g satisfies the homogeneous equation

$$\mathbf{L}^* g = 0 \tag{12.8}$$

throughout $\mathscr{R}$, together with boundary conditions which are such that the combination $U + g$ does satisfy the boundary conditions imposed on G. We

[38] Of course, the ξ_j's need not all be *space* variables.

[39] Note that we avoid saying "... together with *adjoint* boundary conditions" since the boundary conditions on u may be inhomogeneous, in which case the adjoint operator $\mathscr{L}^*$ is *not defined*.

[40] In exceptional cases the "ordinary" Green's function may not exist, and suitable modification of the method is then required (e.g., Exercise 6.9).

have seen that the *principal solution* U is not unique since any function f satisfying $\mathbf{L}^* f = 0$ can be added. In the case of the wave equation, we uncovered both *forward* and *backward running* principal solutions. We saw that the initial-value nature of the problem dictated the selection of the backward running principal solution.

Calculation of U is simplest in the event that $\mathbf{L}^*$ is such that U can be taken to be a function of the single variable

$$r = \sqrt{(\xi_1 - x_1)^2 + \cdots + (\xi_n - x_n)^2} \tag{12.9}$$

since (12.7) then reduces to an *ordinary* differential equation. Otherwise, an *integral transform* line of approach may be best. Sometimes, *physical reasoning* may provide the most direct approach (Exercises 11.1 and 5.1).

If we are lucky, the regular part g can be obtained by inspection, using the method of *images*. If $\mathbf{L}$ is the two-dimensional Laplacian, the method of *conformal mapping* is available. Otherwise, *separation of variables*, *eigenfunction expansions*, or *integral transforms* may be suitable.

Suggested Reading

We highly recommend the following list of related reading. It is by no means intended to be complete.

Bergman, Stefan, and M. Schiffer, *Kernel Functions and Elliptic Differential Equations in Mathematical Physics*, Academic Press, Inc., New York, 1953.

A research monograph, not a textbook. Much information on principal solutions and Green's functions in connection with the partial differential equations of heat conduction, hydrodynamics, electrostatics, magnetostatics and elasticity.

Carrier, George F., Max Krook, and Carl E. Pearson, *Functions of a Complex Variable*, McGraw-Hill Book Company, New York, 1966.

Excellent! Beyond complex variable theory, it includes chapters on special functions, asymptotic methods, integral transforms, and special techniques.

Dennery, Philippe, and André Krzywicki, *Mathematics for Physicists*, Harper & Row, New York, 1967.

Very nice general reference. Includes complex variable theory, linear operators, differential equations, and Green's functions.

Friedman, Bernard, *Principles and Techniques of Applied Mathematics*, John Wiley & Sons, Inc., New York, 1957.

Detailed treatment of spectral theory of differential equations, including Green's functions. Difficult reading unless read from the beginning.

Hildebrand, F. B., *Advanced Calculus for Engineers*, Prentice-Hall, Inc., Englewood Cliffs, N. J., 1948.

Good general reference. Written clearly, at a lower level of difficulty than the above books. Pertinent coverage includes Sturm-Liouville theory, vector integral theorems, Jacobians, partial differential equations, and complex variables.

Jackson, J. D., *Classical Electrodynamics*, John Wiley & Sons, Inc., New, York, 1962.

Classical reference for electrostatics, electrodynamics, including applications of Green's functions.

Lighthill, M. J., *Introduction to Fourier Analysis and Generalized Functions*, Cambridge University Press, New York, 1958.

Concise mathematical discussion of generalized functions.

Mackie, A. G., *Boundary Value Problems*, Hafner Publishing Company, New York, 1965.

Clear, concise account of boundary value problems and Green's functions.

McLachlan, N. W., *Bessel Functions for Engineers*, 2nd ed., Oxford at the Clarendon Press, 1955.

Theory is supplemented by many engineering applications; a helpful list of formulas appears in the Appendix.

Morse, Philip M., and Herman Feshbach, *Methods of Theoretical Physics*, Parts I and II, McGraw-Hill Book Company, New York, 1953.

Encyclopedic work on mathematical methods of physics. Not always easy reading, however.

Sommerfeld, Arnold, *Partial Differential Equations in Physics*, Academic Press, Inc., New York, 1949.

Beautifully written. See, especially, the material on the differences among hyperbolic, parabolic, and elliptic differential equations, and on Green's functions.

Index

A

B

C

D

E

F

G

H

I

J

K

L

M

N

O

P

R

S

T

V

W